PHYSICS RESEARCH AND TECHNOLOGY

A CLOSER LOOK AT MAGNETIC ANISOTROPY

PHYSICS RESEARCH AND TECHNOLOGY

Additional books and e-books in this series can be found on Nova's website under the Series tab.

PHYSICS RESEARCH AND TECHNOLOGY

A CLOSER LOOK AT MAGNETIC ANISOTROPY

GEORGES FREMONT
EDITOR

Copyright © 2020 by Nova Science Publishers, Inc.

All rights reserved. No part of this book may be reproduced, stored in a retrieval system or transmitted in any form or by any means: electronic, electrostatic, magnetic, tape, mechanical photocopying, recording or otherwise without the written permission of the Publisher.

We have partnered with Copyright Clearance Center to make it easy for you to obtain permissions to reuse content from this publication. Simply navigate to this publication's page on Nova's website and locate the "Get Permission" button below the title description. This button is linked directly to the title's permission page on copyright.com. Alternatively, you can visit copyright.com and search by title, ISBN, or ISSN.

For further questions about using the service on copyright.com, please contact:
Copyright Clearance Center
Phone: +1-(978) 750-8400   Fax: +1-(978) 750-4470   E-mail: info@copyright.com.

NOTICE TO THE READER

The Publisher has taken reasonable care in the preparation of this book, but makes no expressed or implied warranty of any kind and assumes no responsibility for any errors or omissions. No liability is assumed for incidental or consequential damages in connection with or arising out of information contained in this book. The Publisher shall not be liable for any special, consequential, or exemplary damages resulting, in whole or in part, from the readers' use of, or reliance upon, this material. Any parts of this book based on government reports are so indicated and copyright is claimed for those parts to the extent applicable to compilations of such works.

Independent verification should be sought for any data, advice or recommendations contained in this book. In addition, no responsibility is assumed by the Publisher for any injury and/or damage to persons or property arising from any methods, products, instructions, ideas or otherwise contained in this publication.

This publication is designed to provide accurate and authoritative information with regard to the subject matter covered herein. It is sold with the clear understanding that the Publisher is not engaged in rendering legal or any other professional services. If legal or any other expert assistance is required, the services of a competent person should be sought. FROM A DECLARATION OF PARTICIPANTS JOINTLY ADOPTED BY A COMMITTEE OF THE AMERICAN BAR ASSOCIATION AND A COMMITTEE OF PUBLISHERS.

Additional color graphics may be available in the e-book version of this book.

Library of Congress Cataloging-in-Publication Data

ISBN: 978-1-53617-566-0

Published by Nova Science Publishers, Inc. † New York

CONTENTS

Preface		vii
Chapter 1	Anisotropy of Assemblies of Densely Packed Ferromagnetic Nanoparticles Embedded in Carbon Nanotubes *Serghej L. Prischepa, Alexander L. Danilyuk and Ivan V. Komissarov*	1
Chapter 2	Enhancement of Perpendicular Magnetic Anisotropy in FeCoZr-CaF$_2$ Nanocomposite Films by Combined Influence of Nanoparticles Oxidation and Ion Irradiation *Julia Kasiuk, Julia Fedotova, Vadim Bayev, Janusz Przewoźnik, Czesław Kapusta, Vladimir Skuratov, Momir Milosavlievič and Jacques O'Connell*	47
Chapter 3	Colloidal Magnetic Fluids: A Special Case of Magnetic Anisotropy *Daniel Mayer and Petr Polcar*	89
Index		141

PREFACE

A Closer Look at Magnetic Anisotropy reports on results related to the impact of magnetic anisotropy on the properties of a new type of nanocomposite consisting of ferromagnetic nanoparticles embedded in carbon nanotubes. The authors demonstrate that when each nanotube contains only one ferromagnetic nanoparticle, the magneto-elastic anisotropy leads to the formation of densely packed arrays of magnetically isolated nanoparticles.

Following this, the aspects of perpendicular magnetic anisotropy of FeCoZr-CaF2 nanocomposite films induced by shape anisotropy of metallic nanoparticles are explored, as well as the methods of anisotropy enhancement by the films treatment.

Several examples of the prospective technical applications of magnetic liquids are presented. Three examples are discussed more in detail: controlled torsion dampers for applications in transportation, ferrofluid controlled capacitors for applications in sensor technology, and peristaltic pumps that take advantage of magneto-elastic properties.

Chapter 1 – The authors report the results related to the impact of magnetic anisotropy on the properties of a new type of nanocomposite consisting of ferromagnetic nanoparticles (NP) embedded in carbon nanotubes (CNT). Samples were synthesized by chemical vapor deposition. The authors found that for low NP concentration, NPs are intercalated mainly inside CNTs and the extended magnetic order, up to hundreds of nanometers, presents in samples. It is shown by analyzing the correlation functions of the magnetic anisotropy axes that the extended order is not simply due to random anisotropy but is associated with the coherent magnetic anisotropy. With increasing temperature, the extended magnetic order is lost, the exchange coupling becomes stronger, but the coherent anisotropy still occurs. The magnetic coupling between NPs distant from each other for tens and hundreds of nanometers could occur via the RKKY interaction. The magnetic relaxation measurements confirmed the importance of the magnetic anisotropy at low temperatures. For the first time, the authors have been able to analyze the relaxation data using the temperature dependence of the magnetic anisotropy. The authors demonstrate further that, when each nanotube contains only one ferromagnetic NP, the magnetoelastic anisotropy plays an important role and leads to the formation of densely packed array of magnetically isolated nanoparticles.

Chapter 2 - The study presented is focused on the aspects of perpendicular magnetic anisotropy in FeCoZr-CaF$_2$ nanocomposite films induced by shape anisotropy of metallic nanoparticles. It also concers the methods of anisotropy enhancement through the films treatment by means of oxidation of metallic nanoparticles and heavy ions irradiation of nanocomposites, as well as the combination of both procedures. Oxidation and irradiation conditions, i.e., the oxygen pressure P_O during films sputtering and Xe ions fluence D, are under consideration for maximal magnetic anisotropy increase. The influence of oxidation and heavy ions bombardment on the crystalline structure, phase composition and magnetic state of nanoparticles as well as on magnetic parameters of the composite films (in particular, the anisotropy and demagnetizing fields) characterizing their anisotropic properties are studied by transmission electron microscopy, X-ray diffraction, Mössbauer spectroscopy and magnetometry. A considerable enhancement of the perpendicular magnetic anisotropy is found in the case of "ferromagnetic α-FeCo(Zr) core – antiferromagnetic α-Fe$_2$O$_3$ shell" structure of nanoparticles formed at $P_O = 4.3$ mPa. It corresponds to more than two-fold increase in the films anisotropy field (up to 3.5 kOe), as compared with that containing non-oxidised ferromagnetic nanoparticles and to a decrease in the the canting angle of their magnetic moments (from 24 to 16°) with respect to the direction of the film normal. Nonmagnetic channels formation is believed to be responsible for the magnetic nanoparticles separation and for the corresponding decrease in the films demagnetizing field as the result of their irradiation by Xe ions with moderate fluences (less than $1 \cdot 10^{13}$ ion/cm^2) that provides increase of perpendicular component of magnetic anisotropy. Protective function of the oxide shells is revealed in the case of high Xe fluence application (up to $2.5 \cdot 10^{13}$ ion/cm^2) that prevents destruction of ordered nanoparticles. Improvement of nanoparticles crystallinity is found in the case of irradiation of the films with complex "core-shell" structure of nanoparticles and is accompanied by the preservation of high perpendicular magnetic anisotropy.

Chapter 3 - Magnetic fluids (thus nanacomposite magnetic material) represent relatively innovative and perspective material for many industrial applications. Recent rapid development of nanotechnologies allowed production of magnetic fluids with wide range of required physical and chemical properties. One of characteristic attributes of magnetic fluids is the significant change of their physical properties in dependence on the application of the external magnetic field. In technical practice, the change of viscosity of these fluids is most commonly used, but other physical properties (e.g., magnetic permeability and dielectric permittivity) change as well. Magnetic fluids are significantly non linear and strongly anisotropic medium. This must be respected when designing different applications by the use of special mathematical apparatus. From the electromagnetic field point of view, the characteristic parameter of the magnetic fluid is the tensor of the magnetic permeability/dielectric permittivity. Presented work describes and innovate method of determination elements of this tensor by measurement carried out on the sample of magnetic fluid. Further, the mathematic-physical properties of magnetic fluid as the medium with and orthogonal anisotropy during the time-varying magnetization are examined. Dissipative phenomena manifesting themselves as energy losses, which significantly impact the design of various electrical appliances, are then discussed. Moreover, new levitation phenomena in magnetic fluids are discussed. The possibility to levitate bodies in magnetic liquid can lead to significant practical applications. Several examples of perspective technical applications of magnetic liquids are presented as well. Three examples are discussed more in detail: controlled torsion damper for application in transportation, next ferrofluid controlled capacitor for application in sensor technology and peristaltic pump that takes advantage of magneto-elastic properties and shows some significant advantages compared to classical pumps design.

In: A Closer Look at Magnetic Anisotropy ISBN: 978-1-53617-566-0
Editor: Georges Fremont © 2020 Nova Science Publishers, Inc.

Chapter 1

ANISOTROPY OF ASSEMBLIES OF DENSELY PACKED FERROMAGNETIC NANOPARTICLES EMBEDDED IN CARBON NANOTUBES

Serghej L. Prischepa[1,2,*]*, Alexander L. Danilyuk*[1] *and Ivan V. Komissarov*[1,2]

[1]Belarusian State University of Informatics and Radioelectronics, Minsk, Belarus
[2]National Research Nuclear University (MEPhI), Moscow, Russia

ABSTRACT

We report the results related to the impact of magnetic anisotropy on the properties of a new type of nanocomposite consisting of ferromagnetic nanoparticles (NP) embedded in carbon nanotubes (CNT). Samples were synthesized by chemical vapor deposition. We found that for low NP concentration, NPs are intercalated mainly inside CNTs and the extended magnetic order, up to hundreds of nanometers, presents in samples. It is shown by analyzing the correlation functions of the

---

[*] Corresponding Author's E-mail: prischepa@bsuir.by.

magnetic anisotropy axes that the extended order is not simply due to random anisotropy but is associated with the coherent magnetic anisotropy. With increasing temperature, the extended magnetic order is lost, the exchange coupling becomes stronger, but the coherent anisotropy still occurs. The magnetic coupling between NPs distant from each other for tens and hundreds of nanometers could occur via the RKKY interaction. The magnetic relaxation measurements confirmed the importance of the magnetic anisotropy at low temperatures. For the first time, we have been able to analyze the relaxation data using the temperature dependence of the magnetic anisotropy. We demonstrate further that, when each nanotube contains only one ferromagnetic NP, the magnetoelastic anisotropy plays an important role and leads to the formation of densely packed array of magnetically isolated nanoparticles.

Keywords: carbon nanotubes, correlation function of the magnetic anisotropy axes, law of the approach to magnetic saturation, magnetic relaxation, magnetoelastic anisotropy

INTRODUCTION

Magnetic nanostructured composites (nanocomposites) consisting of densely packed ferromagnetic nanoparticles (NPs) embedded in matrix material represent an increased performance of devices together with a sustained interest in understanding their fundamental properties. Improved coercivity H_C, saturation magnetization M_S and magnetic anisotropy have been already reported for different kinds of magnetic nanocomposites [1–3]. These magnetic properties are determined by the properties of NPs embedded into matrix material. Single domain NPs are characterized by single magnetic moment with a direction adjusted by local anisotropy. Their stability of the magnetization with time depends on the relation between thermal energy and total anisotropy energy of the NP, KV_{NP}, where K is the effective anisotropy energy density and V_{NP} is the volume of NP. With a noticeable decrease in V_{NP}, the contribution of K to this product should increase in order to maintain thermal stability. The K is generally the superposition of magnetocrystalline (K_{MC}), shape (K_S) and magnetoelastic (K_{ME}) energies. If K_{MC} does not exceed $10^4 - 10^5$ J/m^3 for

3d metals (Fe, Ni, Co), the K_S, which is proportional to the square of the saturation magnetization M_S^2, could reach values of 10^6 J/m^3. Finally, in nanostructured materials plastic deformations are constrained by surfaces and interfaces. As a result, these materials may have significant elastic stresses. The contribution of K_{ME} becomes decisive if elastic stresses of the order of 1-10 GPa are reached. In addition, cooperative effects between ferromagnetic NPs and matrix material could be of interest both from the fundamental and practical point of view. Such kind of interaction is determined by many factors, like arrangement of NPs in the matrix, electrophysical and magnetic properties of the matrix material, as well as NP/matrix interfaces. These factors, in turn, depend on the technology which is applied for the NPs embedding into the matrix. Different approaches are currently used to fabricate matrix dispersed NPs. Among others it is worth mentioning co-precipitation, thermal decomposition, emulsion methods [4], co-evaporation [5], electrochemical processes [6] and chemical vapor deposition (CVD) [7]. The material of the matrix could also vary in a wide range, covering polymers [5], silica [4], porous alumina [8], carbon nanotubes (CNT) [6, 9], etc.

Among all possibilities, CNT-based magnetic nanocomposites are of special interest. Such material presents a porous conducting discontinuous medium. Ferromagnetic NPs can be localized inside or outside of the nanotube inner channels. The aspect ratio of inclusions can vary in a wide range, from 1 to approximately 100, indicating formation of objects with variable morphology, from nanoparticles to nanowires. Number of NPs inside nanotubes can be varied between 1 and many, depending on the fabrication technology.

The role of the CNT in the interparticle interaction is of great importance. Despite porosity, the CNT matrix usually is a well-conducting object. Therefore, this conducting medium can provide indirect exchange coupling (IEC) between ferromagnetic nanoparticles [10, 11] with long-range character [12, 13]. Moreover, NPs inside CNTs can experience significant stresses which reflects on the magnetic anisotropy. All this is reflected in the particular interparticle interaction, creating peculiarities in the interplay between the magnetic anisotropy and exchange coupling [14,

15], which could be studied within the random anisotropy model (RAM) [14, 16, 17]. The contribution of the magnetic anisotropy in the total energy of the system can be so large that it exceeds the dipole interaction. This leads to the formation of ensemble of closely packed magnetically isolated nanoparticles which is important for magnetic data storage. In addition, large values of magnetic anisotropy could influence the magnetization relaxation at low temperatures.

CORRELATION FUNCTIONS OF THE MAGNETIC ANISOTROPY AXES

The mechanisms of the interparticle interaction as well as the magnetic anisotropy contribution can be studied using different methods. One of the most affordable is the investigation of isothermal magnetic hysteresis loops, $M(H)$. Such magnetostatic parameters like H_C, M_S, remanence M_{rem}, and their temperature dependencies can be obtained from the $M(H)$ results. Within such an approach it is possible to derive useful information regarding such magnetic properties of the studied materials like magnetization reversal mechanisms, type of magnetic anisotropy, etc [1, 18, 19]. The shape of the high field part of $M(H)$ curve could give additional important information about the interparticle interaction. Analyzing the law of the approach to magnetic saturation (LAS), it is possible to distinguish the cases of magnetically isolated and strongly interacting via exchange coupling NPs [20–23]. The approach towards magnetic saturation is of specific interest for CNT-based nanocomposites with ferromagnetic NPs because this part of the $M(H)$ curve is defined by general characteristics of a sample and does not contain metastable states [24]. However, the approach developed for the LAS in the past does not take account of possible matrix material contribution, which can significantly vary both the magnetic correlation lengths and the type of coupling between NPs embedded in the magnetic matrix. Even new magnetically ordered states could be induced by peculiarities of interparticle interactions [25, 26].

Usually the exchange coupling between crystalline NPs can be considered when the inequality $R_c < R_f$ is fulfilled. Here R_c is the average radius of the NP and R_f is the exchange correlation length. The peculiarity of the exchange correlations depends on the dimensionality of the NP arrangement. For H far from H_{ex} (the exchange field) phenomenological LAS can be estimated as

$$\frac{\delta M(H)}{M_S} = \frac{M(H) - M_S}{M_S} \sim H^{-\alpha} \qquad (1)$$

where the exponent α depends on the dimensionality d, $\alpha = (4-d)/2$ [16]. For example, for the three-dimensional (3D) case $\alpha = 1/2$, while for the two-dimensional (2D) interaction $\alpha = 1$. When $\alpha = 2$, it means that the magnetic anisotropy dominates. Usually this occurs at $H \gg H_{ex}$. From Eq. (1) it follows that the analysis of the magnetization curve $M(H)$ in the region where $M \to M_S$ could give, in a relatively simple way, useful information about the dimensionality of the NPs interaction which, in turn, could be related to their arrangement. Indeed, many experimental works have been dedicated to such kind of study. In particular, 3D dimensionality was unambiguously established in different amorphous compounds [27] and nanostructured Fe and Ni materials [28].

2D dimensionality is a characteristic feature of thin films with thickness $d_f < R_f$ [29] and of nanocrystalline thick materials ($d_f \gg R_f$), but with anisotropic ferromagnetic correlation lengths, $R_f^{\parallel} \gg R_f^{\perp}$ [30]. Finally, a 1D system of exchange-coupled ferromagnetic NPs was reported for Fe nanowires embedded inside inner channels of CNTs [31]. In all these studies authors usually demonstrated, as the evidence of dimensionality, the $\delta M/M_s$ vs. $H^{-\alpha}$ plot only for a single exponent α. The obtained agreement between the experimental data and Eq. (1) with a single exponent α was used as the main argument, supporting a particular dimensionality.

On the other hand, for CNT based nanocomposites, such elaboration of the experimental data does not lead to the unambiguous statement about the actual dimensionality [14]. It should be noted that the simple relation (1) is not applicable in the range of magnetic fields comparable to the field of exchange interaction. However, usually the measurements of the LAS are carried out precisely when the condition $H \sim H_{ex}$ is met. Consequently, Eq. (1) is no longer valid, the contribution of the magnetic anisotropy becomes significant and a more general approach should be considered, in which the magnetic ordering is described by the correlation function $C(r)$ of the magnetic anisotropy axes in the real space [23]. This approach allows obtaining important information about the mechanism of magnetic interaction between NPs intercalated inside CNT: interplay between exchange coupling and magnetic anisotropy, role of the coherent anisotropy (CA) and its relation with the random anisotropy, and, finally, indication of the impact of the CNT medium on the interparticle interaction [32]. The expression for LAS in this case depends on the dimensionality of the system.

For 1D and 3D the general expression for LAS is [24]

$$\frac{\delta M(H)}{M_S} = \frac{1}{30}\left(\frac{H_{ra}}{H_{ex}}\right)^2 \left(\frac{H_{ex}}{H}\right)^{\frac{1}{2}} \int_0^\infty d^3x\, C(x) x^2 \exp\left[-x\left(\frac{H}{H_{ex}}\right)^{1/2}\right] \quad (2)$$

where H_{ra} is the random anisotropy field, the coordinate x is normalized to the R_a value, the length over which magnetic anisotropy axes are correlated.

For 2D the magnetization approaches saturation as [14]

$$\frac{\delta M(H)}{M_S} = \frac{1}{32}\left(\frac{H_{ra}}{H_{ex}}\right)^2 \left(\frac{H_{ex}}{H}\right)^{\frac{1}{2}} \int_0^\infty d^3x\, C(x) x^2 K_1\left[x\left(\frac{H}{H_{ex}}\right)^{\frac{1}{2}}\right], \quad (3)$$

where K_1 is the modified Hankel function.

The boundary conditions for the correlation function are

$$C(0) = 1, C(x \gg 1) \to 0 \quad (4)$$

The analysis of the experimental data at $H \sim H_{ex}$ according to Eqs. (2) and (3) with boundary conditions (4) provide an explicit form of $C(x)$. Before performing these analysis it is necessary to establish the dimensionalty of the system. This can be done applying the law (1) to the experimental data.

It is worth noticing that in the above analysis the possible contribution of the CA is neglected. Taking account of the CA results in the modification of the LAS. In particular, such auxiliary quantities like the coherent anisotropy correlation length $\delta_{ca} \sim R_a(H_{ex}/H_{ca})^{1/2}$ and field $H_{sa} \equiv H_{ra}^4/H_{ex}^3$ are introduced in the theory [33]. The field of the CA can be evaluated from $C(x)$. In the presence of a field $H > H_{ca}$ in equations for correlation function one needs simply to replace external magnetic field H by $H + H_{ca}$ [33]. Thus the final expression for the LAS in the presence of both random and coherent anisotropy contributions is

$$\frac{\delta M(H)}{M_S} = \frac{1}{30}\left(\frac{H_{ra}}{H_{ex}}\right)^2 \left(\frac{H_{ex}}{H+H_{ca}}\right)^{\frac{1}{2}} \int_0^\infty d^3x\, C(x) x^2 \exp\left[-x\left(\frac{H+H_{ca}}{H_{ex}}\right)^{1/2}\right] \quad (5)$$

The explicit form of the correlation function is obtained by fitting the experimental data and estimating the Laplace integral

$$F_{3D}(p) = 4\pi \int_0^\infty dx \, \exp(-px) x^2 C(x) \quad (6)$$

The correlation function in the 2D case is obtained by fitting the experimental data within the Eq. (3) and applying the K-transform [34],

$$F_{2D}(p) = \int_0^\infty dx \, f(x) (px)^{1/2} K_1(px) \quad (7)$$

Note that the Eq. (3) can be rewritten in a form more convenient for the experimental data elaboration,

$$\frac{\beta}{p^2} = \left(\frac{H_{ra}}{H_{ex}}\right)^2 \frac{4\pi}{32 p^{\frac{3}{2}}} \int_0^\infty dx\, C(x) x^{\frac{3}{2}} (px)^{1/2} K_1(px) \quad (8)$$

With $p = (\frac{H}{H_{ex}})^{1/2}$ and β a constant determined by the slope of the dependence (1).

SAMPLES

Samples studied in this work were fabricated by two different methods. The first one was a floating catalyst chemical vapor depostion (FCCVD). Another one was based on the plasma enhanced CVD (PECVD) on flat substrate on which NPs of Co were preliminary deposited. In both approaches nanotubes were vertically aligned with the packing density $10^{10} - 10^{11}$ cm^{-2}. This quantity determined the density of arrangement of ferromagnetic nanoparticles.

FCCVD was applied for synthesis of vertically oriented densely packed CNTs with iron-based NPs located mainly inside inner channels. This was realized by using small concentration of ferrocene (Fe(C_5H_5)$_2$) as a catalyst [35]. In Figure 1a we show the scanning electron image (SEM) of one of the sample synthesized with ferrocene concentration C_F = 0.6wt%. Transmission electron microscopy (TEM) revealed that iron-based NPs are embedded inside CNTs. This result is presented in Figure 1b. The average distance between NPs located in one nanotube was usually about 100 nm, see Figure 1c. More details about samples preparation by FCCVD and their characterization can be found elsewhere [16, 17, 32].

The second approach is based on the growth of vertically oriented CNT array from fixed catalytic Co NPs preliminary deposited onto Si substrate. A SiO_2 layer of 8 nm thickness was first deposited onto Si(100) substrate. Then Co film of 5 nm thickness was deposited with an evaporation cell in an UHV chamber directly connected to the CVD reactor. Afterwards the substrate was transferred into the UHV CVD reactor, and the metal reduction and the formation of an array of Co NPs were performed by heating up to 973 K at a heat rate of 10 K/min under UHV followed by a final treatment at 973 K in a hydrogen/ammonia mixture at 15 mbar. These conditions were chosen to optimize the formation of homogeneous distribution of NPs. Finally, the metal reduction and the formation of array of Co NPs were performed in thermally activated mixture of hydrogen and ammonia at 973K during 15 min. As a result, a reference sample consisting of an ensemble of Co NPs with an average diameter of 15±5 nm and $n_{NP} \approx 1.2 \times 10^{10}$ cm^{-2} was synthesized on the surface of SiO$_2$/Si substrate.

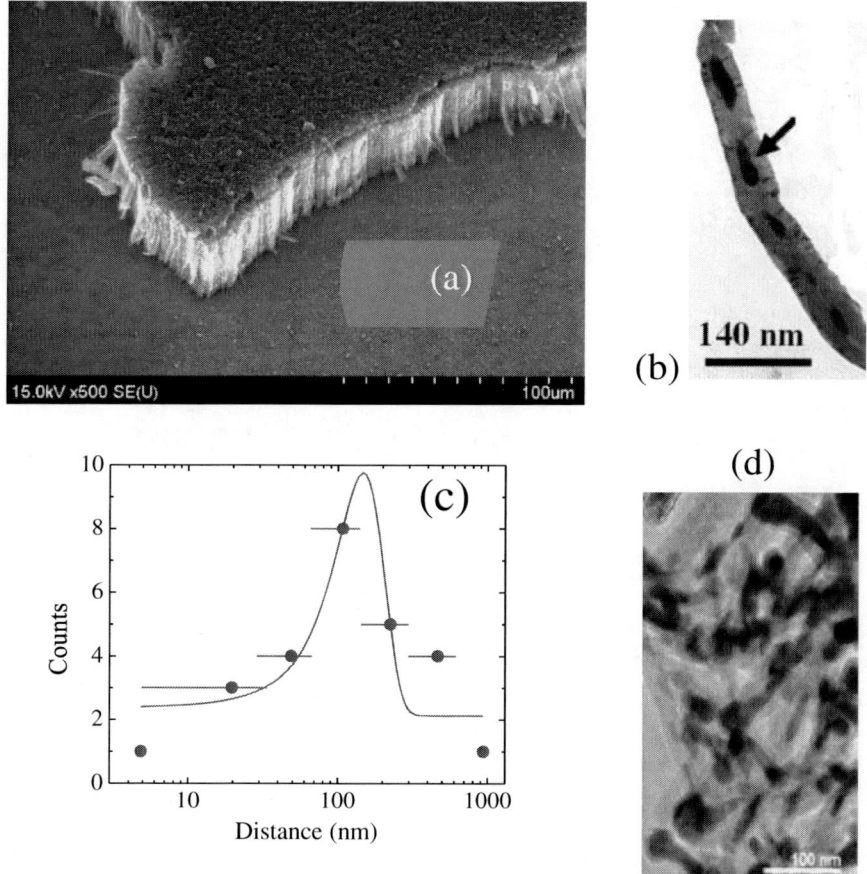

Figure 1. (a) SEM image of vertically oriented CNT array grwon by FCCVD on Si substrate; (b) TEM image of one CNT with ferromagnetic NPs inside (NP is marked by arrow); (c) Average distribution between adjacent NPs along one single nanotube; (d) TEM image of CNTs grown by PECVD with one ferromagnetic NP in each nanotube.

The process of PECVD was performed in the mixture of $C_2H_2:H_2 = 20:80$. In this case, vertically oriented array of CNTs is sytnthesized, each nanotube containes only one ferromagnetic NP located at the top, which is consistent with what is expected for plasma-activated process. The TEM image of such kind of sample is shown in Figure 1d. We emphasize that most of these Co NPs are elongated, as can be seen in Figure 1d, with the long axis oriented parallel to the nanotube axis. The diameter of these inclusions is restricted by the inner CNT channel ($\approx 15 - 20$ nm) and their aspect ratio is about 5, i.e., the NPs have a cylindrical-like shape. Such resulting Co morphology embedded on top of CNT underlines the importance of diffusion processes in/on the NP bulk/surface during the CNT growth. Crystalline structure of Co NPs consists of both cubic *fcc* and hexagonal *hcp* lattice as obtained from SAED study [36]. More details about fabrication of CNT with one ferromagnetic NP by PECVD can be found elsewhere [37-40].

CORRELATION FUNCTIONS OF MAGNETIC ANISOTROPY AXES OF FCCVD SAMPLES

In Figure 2 we analyze the high field part of $M(H)$ curves for sample synthesized by FCCVD with $C_F = 0.6$ wt.%. Data are presented for the temperature range 4-300 K according to the simplified Eq. (1) in order to determine possible dimensionality of the interparticle interaction. Magnetic field is oriented parallel to the CNT axes. In Figs. 2(a)-2(d) we show the correspondence of the experimental data to all possible exponents α in Eq. (1).

It follows from the result of Figure 2 that the dimensionality of the sample can not be determined unambiguously. The main result of this analysis is that at each T it is possible to find a certain range of H, in which one or another exponent α is valid. The situation could be clarified analyzing the width of the magnetic field range, ΔH_α, where the LAS (1) is valid, for each α value and temperature. This result is presented in Figure 3.

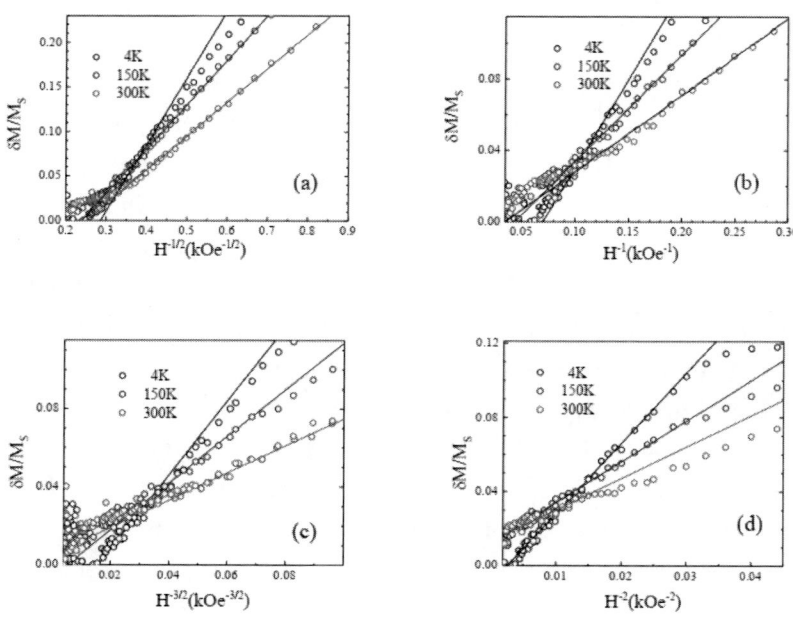

Figure 2. Analysis of LAS at different exponent α values according to Eq. (1) for sample synthesized with $C_F = 0.6$ wt.% by FCCVD. Data are for $T = 4$, 150 and 300K. (a) $\alpha = 1/2$, (b) $\alpha = 1$, (c) $\alpha = 3/2$, (d) $\alpha = 2$.

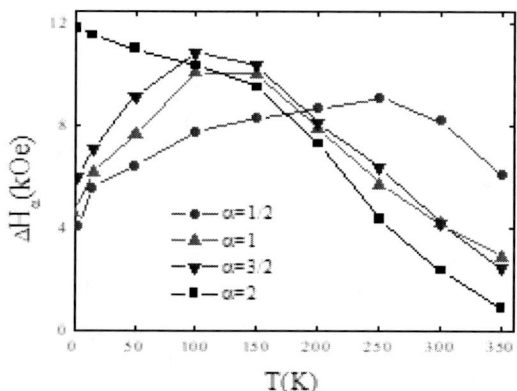

Figure 3. Temperature dependence of ΔH_α obtained from data of Figure 2.

It follows from Figure 3 that the values of ΔH_α vary significantly with the exponent α and T. For the temperature interval 2 – 50 K the widest range of ΔH_α corresponds to $\alpha = 2$. For high temperatures, 200 K < T < 350 K, the exponent $\alpha = ½$ gives the agreement with the experiment in the widest H range. At that interval ΔH_α for other exponents are absorbed by the dominant interval. Finally, for the intermediate temperatures, range between 100 K and 200 K, the ΔH_α values for $\alpha = 2$, 3/2, and 1 are practically the same. Data of Figure 3 are typical for samples with NPs embedded inside CNTs [32]. It means that the dimensionality changes with temperature from mainly 0D at low T to mainly 3D at $T > 200$ K via the mixed state at which different dimensionalities are realized simulteneously.

The quantitative analysis of the experimental data was performed within the standard approach based on the RAM [14, 33, 35, 42, 43] and modified LAS [14, 24, 32, 35, 44-48]. Both these approaches are widely exploited for the analysis of the magnetic properties of amorphous, polycrystalline ferromagnets and CNT-based magnetic nanocomposites.

Within the RAM the local anisotropy field H_{ra} and the exchange field H_{ex} are expressed as

$$H_{ra} = \frac{2K}{M_S} \qquad (9)$$

$$H_{ex} = \frac{2A}{M_S R_a^2} \qquad (10)$$

The constant A is the constant of the exchange coupling. The values of H_{ra} and H_{ex} were evaluated according to the algorithm evolved earlier in [14]. R_a was assumed to be the half of the average diameter of CNT, $R_a \approx R_c \approx \emptyset_{CNT}/2 \approx 15$ nm. Finally, A was calculated as $A = (3 - 5) \times 10^{-12}$ J×m^{-1}. As a result, we get $H_{ra} = 3$–6 kOe and $H_{ex} = 2$–5 kOe. These estimations imply that the values of the exchange field fall within, or are very close to, the range of the LAS. It means that well known simple models for the LAS, strictly speaking, cannot be applied for the quantitative experimental data interpretation, because they have asymptotic character, i.e., are valid only in the limits $H \ll H_{ex}$ and $H \gg H_{ex}$ [16, 41].

In the considered case it is necessary to apply the approach, which operates for the intermediate fields, $H \approx H_{ex}$ [16, 24]. It means that, along with the exchange interaction both local random and coherent anisotropies should be considered [32, 33].

Therefore, we use the modified expression (5) for the LAS analysis. For the best fit procedure of the experimental data the Laplace integral (6) in the T range 2–150 K is close to p^{-3}. It means that, applying the boundary conditions (4) the correlation function is the Fermi-Dirac like

$$C_{0D}(x) = \frac{1}{1+\exp\left(\frac{x-x_{1/2}}{2}\right)} \tag{11}$$

where $x_{1/2}$ is a coordinate at which $C_{3D}(x_{1/2}) = 1/2$. This correlation function is shown in Figure 4 by solid line.

At high temperatures, $T > 200$ K, the Laplace integral was obtained as $F(p) = G\exp(-b/p)p^{-\gamma-1}$, where G, b and γ are constants. For such Laplace representation the correlation function is

$$C_{3D}(x) = x^{\left(\frac{\gamma}{2}\right)-2} b^{-1/2} J_\gamma\left[2(b(x+x_0))^{1/2}\right], \tag{12}$$

where J_γ is the γ-th order Bessel function of the first kind and x_0 is a constant. This correlation function is shown in Figure 4 by dashed line.

In the intermediate T interval (100 K < T <150 K) the experimental data can also be fitted with the Laplace integrals $F(p) \approx p^{-1}$ and $F(p) \approx p^{-2}$. The first Laplace integral leads to the correlation function

$$C_{2D}(x) = \frac{1}{x^2} \cdot \frac{1}{(x^2-a^2)^{\frac{1}{2}}} \left\{ \left[x + (x^2-a^2)^{\frac{1}{2}}\right]^{2m} + \left[x - (x^2-a^2)^{\frac{1}{2}}\right]^{2m} \right\}, x > a \tag{13}$$

where parameters a and m are determined from the fitting procedure. The second Laplace integral gives

$$C_{1D}(x) = \frac{1}{x} \tag{14}$$

The obtained correlation functions are depicted in Figure 4.

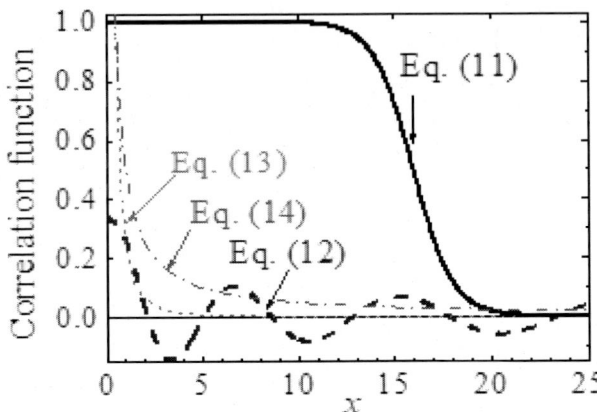

Figure 4. Correlation functions as obtained analyzing the experimental data.

The obtained correlation functions for 2D and 1D dimensionalities thus lead to the following general analytical expression for LAS

$$\frac{\delta M(H)}{M_S} = \frac{4\pi}{15}\left(\frac{H_{ra}}{H+H_{ca}}\right)^2 \left[1 - exp\left(-\frac{H+H_{ca}}{H_{ex}}\right)^{1/2}\right]\left[1 + n \times \left(\frac{H+H_{ca}}{H_{ex}}\right)^{1/2}\right] \tag{15}$$

where the multiplier n is determined from the fitting procedure. In particular, we got that, for $\alpha = 1$ $n = 1$ and for $\alpha = 3/2$ $n = \frac{1}{2}$.

The fitting procedure of the $M(H)$ data in parallel field revealed that the correlation function $C_{0D}(x)$ explains better the experiment at low temperatures. The result of the fitting procedure at $T = 4$ K is shown in Figure 5. In addition, the term of the coherent anisotropy is of great importance. Curve at Figure 5 is plotted with the following parameters: $H_{ex} = 3.8$ kOe, $H_{ra} = 4.0$ kOe, $H_{ca} = 3.5$ kOe. It should be emphasized that without H_{ca} term it was impossible to fit the experimental data. Correlations of the magnetic anisotropy axes in this case propagate up to

150–200 nm. This value is close to the mean distance between adjacent NPs embedded inside CNT, see Figure 1c. Such long-range correlations manifest the crucial role of the coherence anisotropy in the determination of the magnetic properties of CNT-based magnetic nanocomposite.

At high temperatures the $M(H)$ data are fitted better applying the $C_{3D}(x)$ correlation function. The result for $T = 300$ K is shown in Figure 5. Even in this case the fit of the experimental data excluding the mechanism of the coherent anisotropy was impossible. The obtained parameters were as follows, $H_{ex} = 3.0$ kOe, $H_{ra} = 2.3$ kOe, $H_{ca} = 0.8$ kOe.

In Figure 6 we plot the temperature dependence of the evaluated H_{ra} and H_{ca} fields. Shadow areas in Figure 6 mark regions where the mentioned correlation functions are valid. Despite the higher values of the H_{ca} for the $C_{0D}(x)$, the better agreement is reached for the $C_{3D}(x)$ at high T. This means that the crossover from low dimensionality to 3D occurs at temperatures around 200 K. The real H_{ca} value drop significantly due to thermal energy [49].

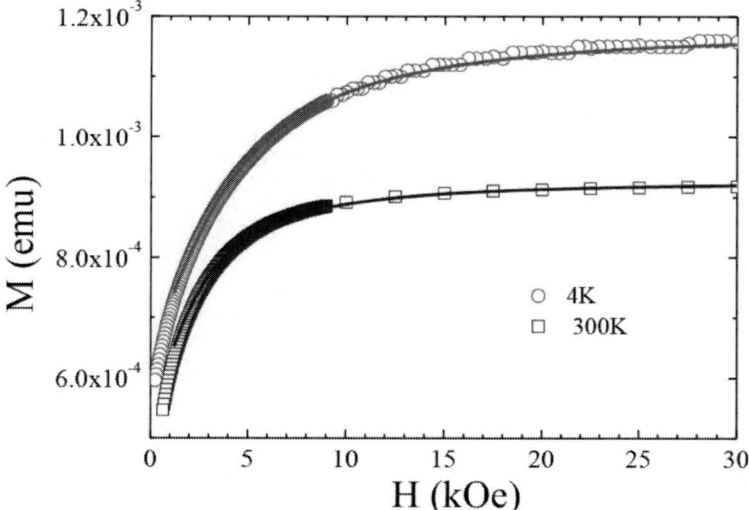

Figure 5. Analyses of LAS at $T = 4$ K and 300 K. Symbols are for the experiment, solid lines refer to the best fit procedure according to Eq. (15) (4 K) and Eq. (5) (300 K). For $T = 4$ K the correlation function was $C_{0D}(x)$, while for $T = 300$ K the correlation function $C_{3D}(x)$ resulted in the best agreement with the experiment.

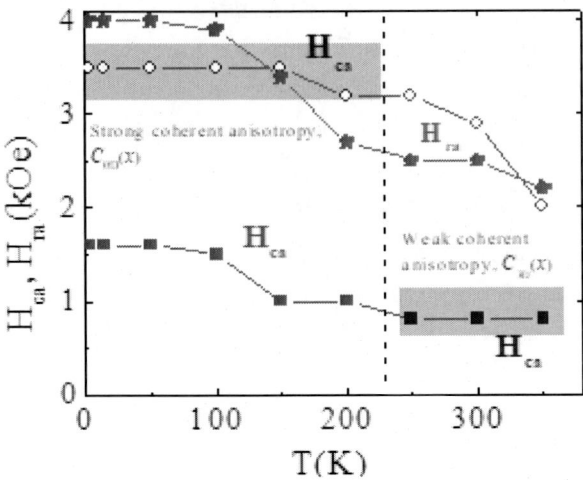

Figure 6. H_{ca} versus T for correlation functions according to the Eq. (11) (upper curve) and Eq. (12) (lower curve). H_{ra} versus T is the same for both functions.

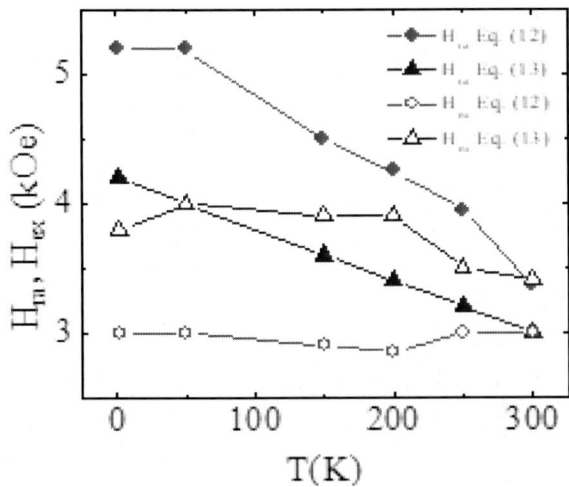

Figure 7. H_{ra} (closed symbols) and H_{ex} (open symbols) versus T extracted from the fitting procedure.

In Figure 7 we present the H_{ra} and H_{ex} versus T as obtained applying two correlation functions, in the 3D and 2D cases. It follows that at low T anisotropy dominates over the exchange interaction between NPs for both $C(x)$. At high T anisotropy still dominates for 3D case, whereas for 2D the exchange interaction is stronger. The latter could be caused by short range character of the correlation function determined by Eq. (13).

The obtained manifestation of the importance of the coherent anisotropy underlines the extremely complex nature of the interparticle interaction via the CNT medium. The alignment of the CNTs also plays a crucial role in the observed long-range correlations of the magnetic anisotropy axes. When the alignment is destroyed, i.e., the oriented array of CNTs becomes a powder, the exponent α in LAS (1) is no longer equal to 2, and becomes equal to ½ [15].

On the other hand, even for strong impact of the coherent anisotropy, the exchange interaction is significant in the system with great (hundreds of nanometers) interparticle distance. Actually, the usual RAM cannot explain this fact. We should involve additional mechanism of indirect exchange coupling (IEC) via the conducting CNT medium. Actually, the peculiarity of the CNT-based nanocomposite is the conductivity of the CNT matrix. In ref. [13] the Rudermann-Kittel-Kasuya-Yosida (RKKY) interaction in single-walled semiconductor CNT with diameters of the order of 1 nm was considered. It was demonstrated that, due to spin-orbit interaction (SOI) [50] the IEC along the CNT axis could reach the value of 1 micrometer [13]. For such a large length of the IEC the constant of spin-orbit splitting equal to 6 meV and the Fermi level shift $\mu \sim 0.5$ eV are required [13, 51]. This significant shift of the Fermi level can be induced by any defects of the nanotubes or their doping [52, 53]. The large spin-orbit coupling was experimentally detected in CNTs [54]. It is reasonably supposed that defective conducting nanotubes investigated in this study could be a suitable environment for the indirect exchange interaction.

Based on the model developed by Klinovaja and Loss [13] we looked for the indications of the possibility of the long-range RKKY interaction in our samples. A theoretical model is based on the consideration of the properties of single-walled CNT (SWNT). The diameter of SWNT does not exceed few nanometers. Strictly speaking, our nanotubes do not fit the criterion of single-walled. However, we can assume with a very effective probability that the IEC occurs through one shell that could be the internal shell. Therefore, we considered for simplicity that the ferromagnetic NP is in contact only with the inner shell of CNT and the coupling propagates along this one inner shell. The conductivity of this shell in defective MWCNTs, as was demonstrated experimentally, could be significant and comparable to the conductivity of the outermost shell [55, 56]. The diameter of the inner shell determines the diameter of the NP and in our calculations was selected as $\varnothing_{CNT} = 25$nm. The axis Z is oriented along the CNT axis. To analyze the decay of the RKKY interaction in the presence of SOI the low frequency component of the oscillations of the normalized spin susceptibility χ along the CNT axis was examined,

$$\chi(z) = \text{si}(2k_+|z|)/2 \pm \text{si}(2k_-|z|) \tag{16}$$

where

$$\text{si}(y) = \int_0^y \frac{\sin t}{t} dt - \frac{\pi}{2}$$

The effective wave vectors k_+ and k_- are defined within the model [13] and depend on the diameter of the CNT, its chirality, parameters of the SOI and μ. For certain combinations of these parameters the exchange coupling is long-range and propagates over hundreds of nanometers [13].

Within this approach, we calculated χ looking for a value of the shift of the Fermi level μ due to SOI, which would lead to the decay length of the order of hundreds of nanometers. It should be noted that μ was the only completely free adjustable parameter. The obtained result is presented in Figure 8a, in which we plot the amplitude envelope of the spin susceptibility oscillations along the z-axis for $\mu = 25$ meV. Other parameters used in these calculations are as follows. Chiral indices were chosen as (235,129). These values correspond to the $\varnothing_{CNT} = 25$nm. The angle of chirality in this case is equal to 39.54° and the circumferential direction $k_G = 0.053$ nm^{-1}. The constant of spin-orbit splitting Δ_{SO} following data of other authors, was chosen to be equal to 6 meV [13, 51]. Finally, the period of oscillations of spin susceptibility for this set of parameters is of the order of 0.5 nm, this result is shown in Figure 8b.

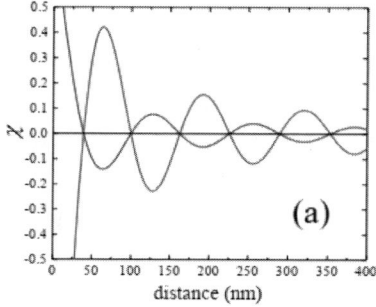
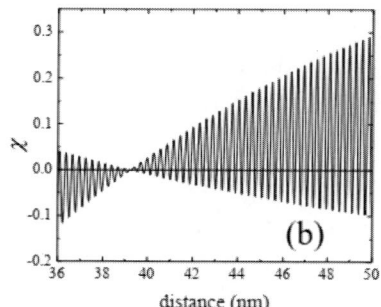

Figure 8. (a) Envelope of the spin susceptibility amplitude oscillations along the CNT axis; (b) High frequency oscillations of the spin susceptibility in the real space.

The obtained chiral indices emphasize the fact that the curvature effects are rather negligible, as should be in a purely theoretical case for a SWCNT of such large diameter. It follows from the result of Figure 8 that for the considered parameters of the nanotube the IEC could propagate on hundreds of nanometers. Even if this result can be regarded only as an indication that the long-range interaction can really occur in our samples, it strongly supports the idea that the IEC directly affects the coupling between ferromagnetic NPs via the CNT medium as was clearly derived from the elaboration of the *M*(*H*) data. To make this statement more solid, it is necessary to carry out the calculations for multi-walled CNT. In addition, the external magnetic field strengthens the SOI and, consequently, could intensify the exchange coupling at large distances.

IMPACT OF THE MAGNETIC ANISOTROPY ON THE RELAXATION OF THE MAGNETIC MOMENTS IN FCCVD SAMPLES

Samples studied in this research have been synthesized by FCCVD with high ferrocene concentration, $C_F = 10$ wt.%. In this case ferromagnetic NPs are localized almost uniformly throughout the sample, i.e., inside and outside CNTs, see Figure 9. The interparticle interaction can be described well within the RAM [14, 16]. Local magnetic moments of single domain NPs are oriented randomly and the exchange interaction between them exceeds the gain in energy due to magnetic anisotropy. Correlations of the magnetic anisotropy axes decay rapidly on a short range, order of the NP size, see Eq. (12). In the light of these results, it is interesting to study the influence of the distribution of orientations of the local magnetization vectors and of the magnetic anisotropy induced barriers between metastable magnetic states on the relaxation processes. This is one of the key points in understanding the mechanisms of interaction of alternating magnetic field with a CNT-based magnetic nanocomposite.

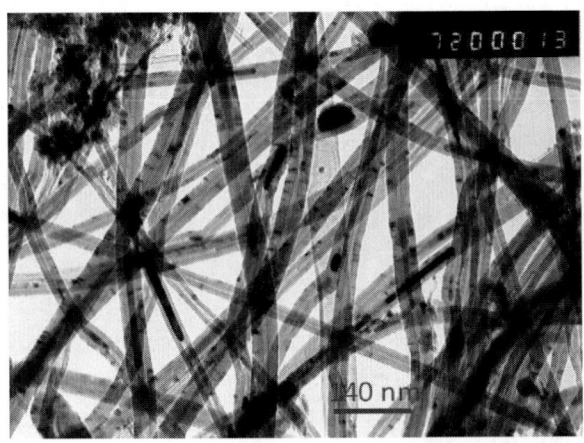

Figure 9. TEM image of CNT-based nanocomposite synthesized with $C_F = 10$ wt.%.

The main phase of the NP material is cementite Fe_3C [14, 57]. Detailed information regarding the structure of carbon matrix can be obtained from Raman spectroscopy. Typical Raman spectrum of studied CNT is presented in Figure 10. The spectrum is dominated by three pronounced bands: one non dispersive band G centered at ~1580 cm^{-1}, and two dispersive bands, D and 2D. In our experiment D and 2D bands are centered at ~1350 cm^{-1} and ~2700 cm^{-1}, respectively. G band is associated with the in plane E_{2g} optical modes, whereas 2D band comes from TO modes and involves two phonons (second order process). Contrary to its second order overtone 2D band, which is always Raman active, D-band requires the presence of defects for its activation [58]. Thus, the intensity of D band is generally accepted as an indicator of defect amount in graphitic systems. Besides, the crystalline size, L_a, can be evaluated from the intensity ratio I_D/I_G [59]. In the case of our sample we get $L_a \sim 32$ nm. In its turn, the intensity of 2D band is proportional to long range order in MWCNT [60]. Moreover, several minor (in terms of intensity) bands are observed in the spectra. The positions of these bands centered at appx. 1100 cm^{-1}, 1470 cm^{-1}, 1620 cm^{-1}, 2450 cm^{-1} and 2900 cm^{-1}, denoted in the plot as D'', D_3, D', D+D'', D+D', respectively. However, the assignment of these peaks is not unambiguous and different interpretation can be found in the literature. For example, the low frequency shoulder of D band which appears at appx. 1100 – 1200 cm^{-1} is often denoted by D_4 [61]. On the other hand, assignment of D+D' to D+G mode [62] also can be found in the literature. The detailed overview of Raman spectroscopy of carbon-based materials with related references can be found elsewhere [63], we only would like to emphasize here that the presence of D_3 and D_4 is the sign of the presence of amorphous carbon and hydrocarbons built-in graphitic matrix [61].

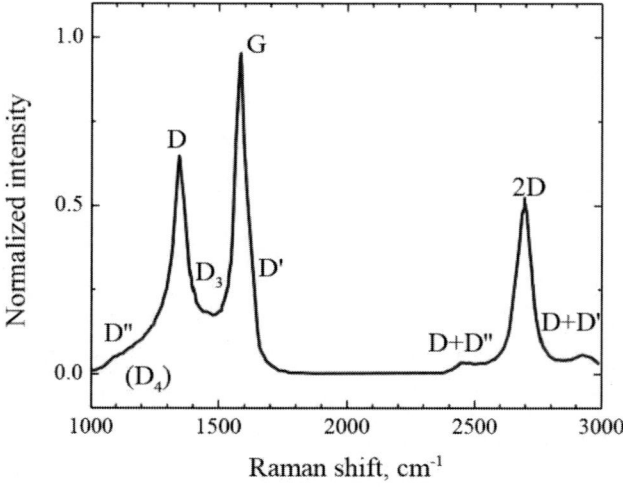

Figure 10. The Raman spectra of synthesized CNTs. The alternative assignment is in the brackets (see the text).

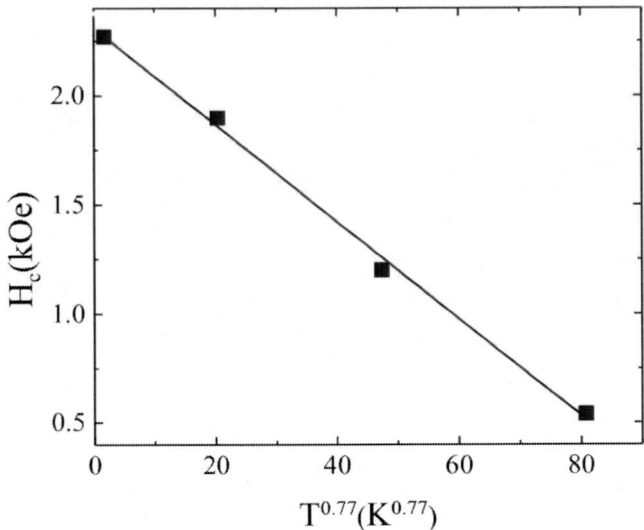

Figure 11. Coercivity versus temperature (2 K < T < 300 K). Solid line is the best linear fit.

Magnetic hysteresis loops were measured for this sample in the temperature range 2 – 300 K. Data were similar to the previously reported in [14, 15]. The coercivity fields H_C were extracted from these measurement. Their values vary in the range from 0.5 to 2.2 kOe, which is much greater than those usually reported for bulk Fe_3C [64]. We believe that this is a clear indication of small size effects on the H_C values, i.e., the coercivity of single domain NP is usually much greater than those of bulk [1]. For this case and for the intermediate temperatures H_C is expected to follow the law expressed in Eq. (17):

$$H_C(T) = H_C(0)\left[1 - (T/T_B)^\zeta\right], \qquad (17)$$

where T_B is blocking temperature of the largest particles and exponent ζ depends on the alignment of the local magnetic moments of particles [1]. In particular, for random orientation $\zeta = 0.77$ [65]. In Figure 11 we show the H_c versus $T^{0.77}$ dependence. The solid line represents the best linear fit. The result of Figure 11 confirms the random distribution of local magnetic moments. Moreover, we have also estimated the blocking temperature of the largest particles: T_B is close to $T = 415$ K. This value is smaller than the Curie temperature of cementite, 481 K.

From the above analysis we may conclude that studied CNT-based nanocomposite should be characterized by random magnetic anisotropy. Such systems are well described within the RAM [42]. The LAS can provide in this case a valuable information about the dimensionality of the system, correlations of magnetic anisotropy axes [32, 33, 46, 47] and the dominant role of the interparticle interaction [14, 35]. For all temperatures, the best fit to the $M(H)$ data was given by the $H^{-1/2}$ law (Figure 12), which corresponds to the 3D exchange coupled system.

The correlation function of the magnetic anisotropy axes was desribed by the Eq. (12). In Figure 13 we show measured $M(H)$ dependences along with the results of the best fit procedure according to Eq. (2). The term of the coherent anisotropy was absent in this case. The exchange field was 2.9 kOe for the whole T range while the H_{ra} varies between 5.4 kOe ($T = 2$ K) and 3.2 kOe ($T = 300$ K).

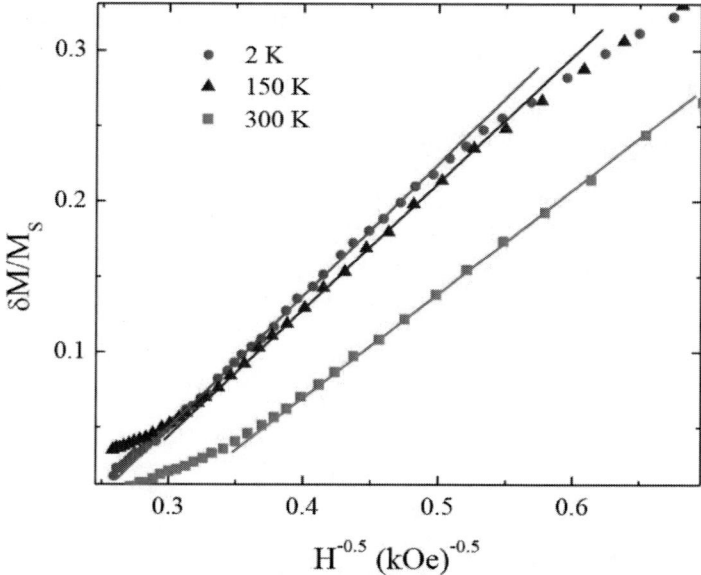

Figure 12. Approaching to magnetic saturation in the temperature range 2 K–300 K for CNT-based magnetic nanocomposite synthesized by FCCVD with $C_F = 10$ wt.% of ferrocene.

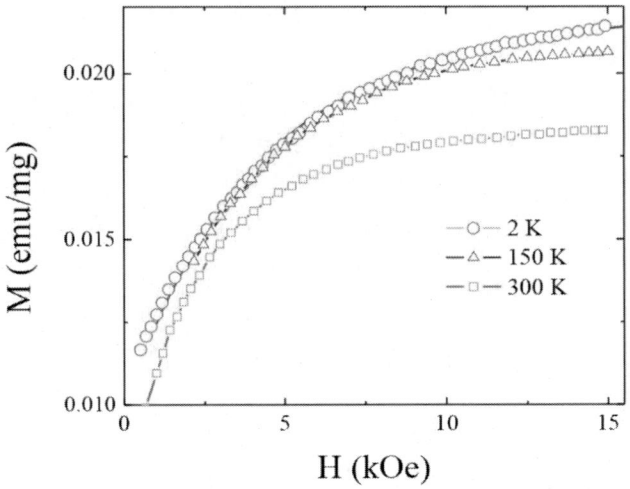

Figure 13. Measured (symbols) and calculated (lines) $M(H)$ dependences for $T = 2$, 150 and 300 K.

On this sample we performed the magnetic relaxation measurements. It was first saturated at $H = 15$ kOe and after we did change the magnetic field to the value of -50 Oe, i.e., it was applied in the opposite direction. Instead of simply remove the magnetic field to zero, the presence of this negative small magnetic field pretends to compensate the possible magnetic remanence of the SQUID magnetometer. This experiment was done in a wide range of temperatures (from 2 K to 300 K) and the magnetization was measured as a function of time. As described in Eq. (18) [66-68] during the relaxation process the magnetization decays in a logarithmic way with time. By studying the slope in this process (i.e., the demagnetization velocity) the characteristic magnetic viscosity (S) can be determined,

$$M(t) = M(t_0)\left[1 - S(T,H)\ln\left(\frac{t}{t_0}\right)\right] \tag{18}$$

where t_0 is the inverse of the attempt frequency. Figure 14a shows the S values calculated for each one of the different temperatures at which measurements were performed. There is a nonlinear increasing tendency of S with respect to temperature which reflects the distribution of the barriers for magnetic anisotropy between various metastable magnetic states. The obtained $S(T)$ dependence makes it possible to restore the distribution function of the magnetic nanoparticles $f(V)$ in the volume V, which is related to the magnitude of the potential barrier in the relaxation of the magnetization. According to the model of slow relaxation at $H \ll H_{ra}$ the magnetic viscosity $S(T,t)$ is expressed as [66-68]

$$S(T,t) = \frac{kTV_B f(V_B)}{K(T)\langle V \rangle \int_0^\infty f(V)dV}, \qquad (19)$$

where k is the Boltzmann constant, $K(T)$ is the temperature dependent anisotropy constant, V_B is the volume of NPs which effectively contribute to the relaxation, $V_B = \frac{kT}{K(T)} ln(t/t_0)$ and $\langle V \rangle = \frac{\int_0^\infty f(V)V dV}{\int_0^\infty f(V)dV}$.

The analysis of Eq. (19), considering the obtained $S(T)$ dependence, Figure 14a, showed that the distribution function can be represented in the form

$$f(V) = \frac{1}{V} exp\left[-\varphi(T)\frac{V}{V_0}\right], \qquad (20)$$

where $\varphi(T)$ is a test function that depends only on temperature, $V_0 = kT_B \ln\left(\frac{t}{t_0}\right)/K_B$ and K_B is the anisotropy constant at $T = T_B$. The presence of the function $\varphi(T)$ is explained by the fact that the parameters affecting the magnetic viscosity (such as the anisotropy constant, the saturation magnetization) depend on temperature in the considered temperature range. In addition, the experimental dependence $S(T)$ is essentially non-linear.

Substituting Eq. (20) into Eq. (19), taking into account that the quantity $\langle V \rangle \int_0^\infty f(V)dV = V_0/\varphi(T)$, we obtain the transcendental equation for the function $\varphi(T)$

$$S(T,t) \ln\left(\frac{t}{t_0}\right) = \varphi(T) \frac{T}{T_B} \frac{K_B}{K(T)} exp\left[-\varphi(T) \frac{T}{T_B} \frac{K_B}{K(T)}\right] \qquad (21)$$

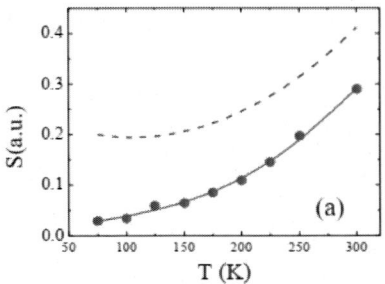 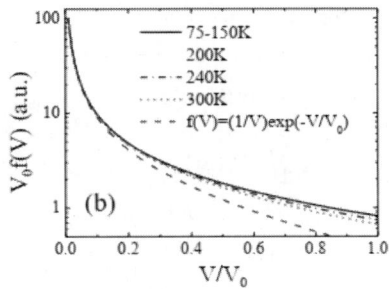

Figure 14. (a) Magnetic viscosity versus temperature with a reverse magnetic field applied of 50 Oe after the saturation (points). Dashed and solid lines correspond to the test function φ(T) and S(T)ln(t/t₀) dependences, respectively; (b) The normalized distribution function f(V) as obtained at different temperatures. Dashed line is for f(V) = (1/V)exp(-V/V₀) distribution function.

The temperature dependent anisotropy constant can be evaluated from RAM [16, 32]

$$[K(T)]^2 = \frac{H_C(T) A^3 M_s(T)}{114(kT_B)^2}, \qquad (22)$$

where A is the exchange constant, the coercive force H_C is defined by the Eq. (17). Considering both the $H_C(T)$ at $T < T_B$ (Figure 11) and the $M_s(T)$ (the Bloch law) dependencies for the considered sample [14], we arrive to the expression

$$S(T,t)\ln\left(\frac{t}{t_0}\right) = F(T)\exp(-F(T)) \qquad (23)$$

where function

$$F(T) = \gamma_0 \varphi(T) \gamma(T), \qquad (24)$$

and

$$\gamma(T) = \frac{T}{T_B}\left\{\left[1 - \left(\frac{T}{T_B}\right)^{0.77}\right](1 - BT^{3/2})\right\}^{-1/2} \qquad (25)$$

In. Eq. (25) B is the Bloch constant, and γ_0 is

$$\gamma_0 = \frac{\sqrt{114 K_B (k T_B)}}{\left(H_C^0 A^3 M_S^0\right)^{1/2}}, \tag{26}$$

where H_C^0 and M_S^0 are the coercivity and saturation magnetization, respectively, at $T = 0$.

With Eq. (23) on the base of the obtained $S(T)$ dependence (Figure 14a) and assuming that $\ln(t/t_0) = 25$, we first find the function $F(T)$. Afterwards, within the Eq. (24) we extract the test function $\varphi(T)$, which after the subsequent approximation can be represented as a polynomial of the second degree

$$\varphi(T) = a + bT + cT^2 \tag{27}$$

For the considered case $a = 0.258$, $b = -1.216 \times 10^{-3}$ K^{-1} and $c = 5.755 \times 10^{-6}$ K^{-2}.

To fit the $S(T)$ dependence we used the previously obtained parameters for such samples as well as measured in this work, $A = (3-5) \times 10^{-12}$ J/m, $M_S^0 = (1.26-1.33) \times 10^6$ A/m, $K_B = 10^4$ J/m^3, $B = 1.65 \times 10^{-5}$ K$^{-3/2}$ [14, 15, 32], $H_C^0 = 2.132$ kOe, $T_B = 415$ K (Figure 11).

The obtained test function $\varphi(T)$ and $S(T)\ln(t/t_0)$ dependences are shown in Figure 14a by dashed and solid lines, respectively. As it follows from this result, a good agreement with the experiment is reached confirming the proposed hypothesis about the distribution function as given in Eq. (20).

It follows from Figure 14a that the test function $\varphi(T)$ in the range 75 – 150 K is temperature independent, whereas the magnetic viscosity $S(T)$ times $\ln(t/t_0)$ is determined by $F(T)$. It means that the $S(T)$ dependence is determined only by $\gamma(T)$, in which the linear increase is modified by the temperature dependence of the anisotropy constant. That is, in the temperature range 75–150 K the linear dependence of the magnetic viscosity is modified by the temperature dependence of the anisotropy constant $K(T)$. This suggests that in this T range the relaxation process is mainly determined by the anisotropy of the sample. In the range 150–300

K the effect of the test function $\varphi(T)$ is manifested. Its influence leads to correction of the function $F(T)$ in the direction of the increase with respect to $\gamma(T)$.

We would like to underline, that in the entire temperature range the magnetic viscosity with high accuracy can be expressed by the expansion

$$S(T)\ln(t/t_0) = F(T)[1-F(T)+(1/2)F^2(T)], \qquad (28)$$

in which terms in square brackets contribute to the region $T > 150$ K. The approximated expression (28) is very convenient for the experimental data fitting instead of Eq. (23).

In addition, we obtained that, at $T > 150$ K the test function tends to compensate the temperature contribution of the anisotropy constant which leads to the practically linear $S(T)$ dependence within the range 230 – 300 K. This is probably because in this temperature range the magnetic viscosity becomes independent on the variation of the anisotropy constant. In other words, with an increase in temperature, when the anisotropy constant decreases and the contribution of thermal fluctuations increases, the relaxation mechanism ceases to react to the anisotropy change.

In Figure 14b we show the distribution function $f(V)$ at different temperatures obtained using the calculated test function $\varphi(T)$. Data are restricted to the region $V < V_0$, which means that the temperature is well below blocking. This is a necessary condition for the application of the slow relaxation model [67]. It follows from Figure 14b that the distribution function is practically independent on temperature. Only for $T = 300$ K the values of the distribution function differ from the values at lower temperatures, when the volume V tends to V_0. This could reflect the fact that at $T = 300$ K we are too close to the blocking temperature and the applicability of the model begins to experience limitations.

The test function $\varphi(T)$ is an important factor in the successful fitting of the magnetic viscosity data. Without this function, the distribution function becomes equal to $(1/V)\exp(-V/V_0)$. With this function we failed to fit the experimental data. In Figure 14b, for comparison, by dashed line we plot the $f(V) = (1/V) \exp(-V/V_0)$ distribution function. It can be seen that it differs from that obtained, especially with increasing volume.

IMPACT OF MAGNETIC ANISOTROPY ON THE MAGNETIC ISOLATION OF CLOSELY PACKED CO NANOPARTICLES EMBEDDED IN CNT GROWN BY PECVD

It is now well known that in order to form a densely packed array of ferromagnetic NPs, prevent their agglomeration and achieve long-time protection against external environment, one of the best ways is to embed them into the carbon nanotubes (CNTs) [69-72]. Selective introducing into the inner channel of multiwalled CNTs could give rise to a new family of hybrid materials with novel functionality which can find applications in magnetoelectronics. The dipole-dipole interaction (DDI) usually dominates for the density of nanosized ferromagnets greater than $n_{NP} \approx 10^9 - 10^{10}$ cm^{-2} [73]. Therefore, to create a system of densely packed magnetically isolated ferromagnetic nanoparticles with $n_{NP} > 10^{10}$ cm^{-2}, it is necessary to increase the contribution of the magnetic anisotropy in the total energy of the system.

Figure 15a presents the MFM image for the reference sample, and Figure 15b shows the MFM result for an array of CNTs with Co on the top of each CNT. The intensity variations are a measurement of the attractive (repulsive) magnetic force gradient.

In the first case, it is possible to distinguish magnetic domains with an average size of about 500 nm encompassing a set of many NPs, which indicates strong DDI between them. For Co-CNT sample, which was grown from a reference one, the average size of magnetic dipole is about 50 nm, which is close to the average size of Co at the top of the CNT. Thus, Co NPs are magnetically separated from each other.

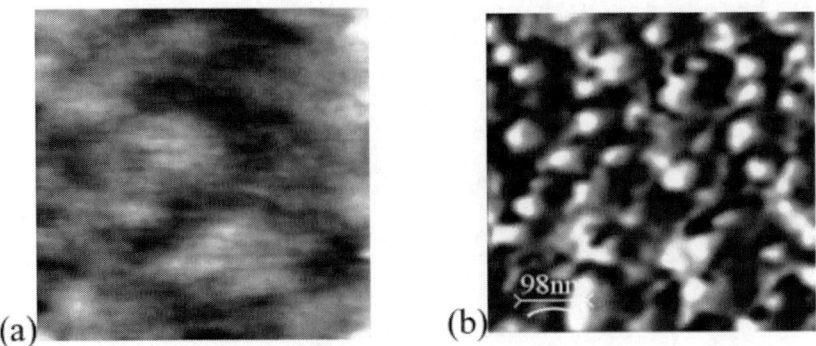

Figure 15. (a) MFM image of Co NPs on SiO$_2$/Si substrate. The size of image is 1.5 μm × 1.5 μm; (b) MFM image of Co NPs embedded inside CNT.

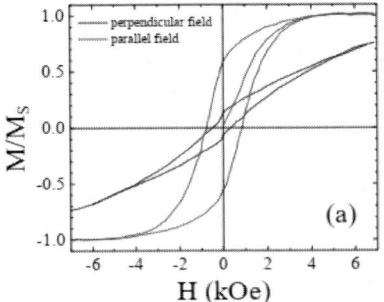
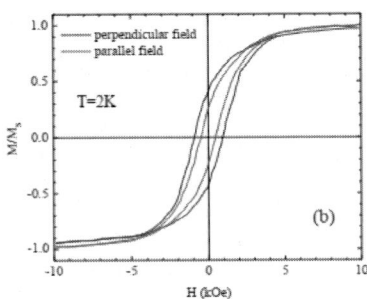

Figure 16. $M(H)$ curves for the parallel (red color) and perpendicular (blue color) magnetic field orientations. $T = 2$ K. (a) reference sample; (b) Co-CNT.

The difference in the magnetic interaction between the particles on the substrate and inside the CNT is also manifested in the shape of hysteresis loops. In Figure 16a the $M(H)$ curves for the reference sample is shown for magnetic field oriented both parallel and perpendicular to the substrate. Rapid saturation in the parallel field along with greater values of the remanence and coercivity can be reasonably attributed to an axis of easy magnetization oriented parallel to the substrate. This can be naturally explained by strong DDI between NPs, because of which the array of NPs can be considered as a thin film.

For Co-CNT sample $M(H)$ loops approach each other for a parallel and perpendicular field directions, see Figure 16b. Usually for an ensemble of Co NPs with the aspect ratio ≥ 3-5 the easy axis is oriented along the cylinder and *M(H)* shapes for the easy and hard axes directions differ significantly [74]. Therefore, most likely, in our Co-CNT samples there is a complex case where magnetocrystalline and, possibly, magnetoelastic anisotropy of NPs have a dispersion relative to some direction at a certain angle to the axis of the nanocylinders. This fact leads to uncertainty in the direction of the axis of easy magnetization, which requires additional analysis.

We start again from the RAM, the validity of which for CNT-based nanocomposites was shown in previous sections of this chapter. The effective magnetic anisotropy constant K which includes contributions of the magnetocrystalline and shape anisotropies, the magnetoelastic contribution and the DDI (K_{DD}), can be evaluated within the random anisotropy model, which has been successfully applied in the past to explain the properties of ferromagnetic NPs in CNT [32],

$$H_C = 114K^2(kT_B)^2/(A^3 M_S). \tag{29}$$

Below the results will be presented for the parallel field. For the perpendicular field the data are similar. The T_B (400 K), H_C (6.53 × 10^4 A/m for reference sample and 3.76 × 10^4 A/m for Co-CNT) and M_S (1.4 × 10^6 A/m) were measured. For nanosized Co we took from literature A = 1.54 × 10^{-11} J/m [75]. Thus, for the reference sample K = 3.77 × 10^5 J/m^3 and for Co-CNT sample K = 2.65 × 10^5 J/m^3.

The contribution of the DDI is expressed as [76]

$$K_{DD} = (1/8\pi)(\mu_0 M_S^2 V_{NP}/\zeta^3)s_d, \tag{30}$$

where ζ is an average interparticle distance, s_d = 6 – 8 is a lattice sum that depends on the type of lattice in the NPs arrangement and the dimensionality of the sample, V_{NP} is the NP volume and μ_0 is magnetic permeability of vacuum. From Eq. (30) we estimate K_{DD} = 1.4×10^5 J/m^3 for the reference sample and K_{DD} = 1.1 × 10^5 J/m^3 for the Co-CNT. The contribution of the SA for the reference sample can be neglected due to the almost equal transverse and longitudinal demagnetizing factors. In addition, we believe that the contribution of the magnetoelastic anisotropy will be unimportant because the Co nanoparticles are in a practically free state.

Therefore, the anisotropy constant K for the reference sample is a function of only K_{DD} and K_{MC}. This results in $K_{MC} = 2.4 \times 10^5$ J/m^3. This value is assumed to be the same for both types of samples. The obtained K_{MC} is slightly greater that K_{DD}. Nevertheless, the impact of the DDI in the presence of only of MCA is determined by the dipolar coupling constant $\alpha_D = K_{DD}/K_{MCA}$ [77]. There is a crucial parameter α_{cr} that determines the transition from a single particle to the collective behavior. The reported values for *fcc* or *hcp* Co NPs are $\alpha_{cr} = (0.2-0.4)(d_n/\zeta)^3$ [78]. If $\alpha_D > \alpha_{cr}$, then DDI dominates, otherwise the MCA prevails. For our ensemble of Co NPs we get $\alpha_{cr} \approx 0.14 < \alpha_D = 0.58$, which indicates that the energy contribution of the DDI is sufficient to form regions covering many NPs, that is indeed observed experimentally.

For the Co-CNT sample, the contribution of the SA and MEA is important. Anisotropy constants K_{SA} and K_{DD} of the system of magnetic nanocylinders can be merged in a single contribution as [79]

$$K_{SA} + K_{DD} = -(1/4)\mu_0 M_S^2 (1 - 3P). \tag{31}$$

Here $P = (\pi r^2 S)$ is the porosity of Co nanowires assembly, $P = 0.06$. From Eq. (31) we get $K_{SA}+K_{DD} = -5.1 \times 10^5$ J/m^3, which leads to $K_{SA} = -6.2 \times 10^5$ J/m^3. Finally, on the base of the above estimations we calculate the value of the magnetoelastic anisotropy constant, $K_{ME} = 5.4 \times 10^5$ J/m^3. This energy density is greater than those reported in literature for Co nanowires embedded in porous matrix [80]. In Table 1 we present the results of the evaluations of the SA, MCA and DDI for both samples for the parallel field. We do not show K_{ME} in this Table. The discussion of this issue will be done further.

Table 1. Evaluations of the SA, MCA and DDI for the parallel orientation of the magnetic field. d_n is an average diameter of the NP and L is an average height of the nanocylinder

Parameter	Unit	Reference sample	Co-CNT
K_{SA}	J/m³	---	-6.16×10^5
K_{DD}	J/m³	1.4×10^5	1.09×10^5
K_{MC}	J/m³	2.37×10^5	2.37×10^5
$K_{SA}+K_{DD}$	J/m³	---	-5.07×10^5
$\alpha_D = K_{DD}/K_{MC}$	a.u.	0.61	---
ζ	nm	32.6	51.5
S	cm⁻²	3×10^{10}	1.2×10^{10}
P	a.u.	0.15	0.06
d_n	nm	25 ± 8	15 ± 5
L	nm	---	100-150

For the Co-CNT array we have to consider both the SA and MEA. Their contribution along with the MCA should exceed the contribution of the dipole interaction between NPs.

The type of Co crystalline lattice affects significantly the interpretation of data on the magnetoelastic anisotropy. For *fcc* Co the K_{ME} can be evaluated as

$$K_{ME}^{fcc} \approx -(3/2)\lambda\sigma, \qquad (32)$$

where λ is the magnetostriction constant and σ is the elastic stress. Using for *fcc* Co $\lambda = -50 \times 10^{-6}$ [81] and applying the obtained K_{ME} value we estimate $\sigma \approx 7.13$ GPa.

For *hcp* Co when the hexagonal axis is parallel to the CNT axis, the K_{ME} is expressed as [82]

$$K_{ME}^{hcp} = \sigma(\lambda_A + \lambda_B)\varepsilon_1, \qquad (33)$$

where λ_A and λ_B are magnetostriction constants and ε_1 is strain. From Eq. (33) we get the tensile stress $\sigma \approx 3.4$ GPa.

For the hexagonal axis oriented perpendicular to the CNT axis the situation changes significantly. The expression for the K_{ME}^{hcp} depends on the orientation of the crystal magnetization $\vec{m}$ [81]. However, the exact values of the direction cosines α_i of $\vec{m}$ with respect to the hexagonal axis are not known in our case. Therefore, for random α_i

$$K_{MEA}^{hcp}(\vec{m}) \approx B_2 \varepsilon_3 + B_3(\varepsilon_1 + \varepsilon_2), \tag{34}$$

where B_i are magnetoelastic coupling coefficients, the indices number the axes of the hexagonal crystal (index 3 corresponds to the hexagonal axis c, indices 1 and 2 correspond to the a and b axes, respectively). We used $B_2 = -29 \times 10^6$ J/m^3, $B_3 = 28.2 \times 10^6$ J/m^3 [81]. The strain ε is found from the system of equations

$$c_{ij}\varepsilon_j = \sigma_i, \tag{35}$$

where c_{ij} are elastic stiffness constants, $\sigma_1 = \sigma_3 = \sigma$ and $\sigma_2 = 0$. For the considered case we evaluate $\sigma = -9.5$ GPa. Such internal elastic stresses cause the strain of the CNTs lattice not more than $\varepsilon \approx 10^{-3}$, which is in good agreement with the literature [83].

The performed evaluations show that the DDI is suppressed mainly by the SA and MEA. The largest contribution of the magnetoelastic anisotropy occurs when the hexagonal axis is oriented parallel to the substrate. This is confirmed by the results of micromagnetic simulation, which was performed for *hcp* Co nanocylinders with diameter 20 nm and length 100 nm. We applied the Nmag package [84] based on the Landau-Lifshitz-Gilbert equation.

Axis Z is oriented along the nanocylinder axis (i.e., the CNT orientation), and XY plane is lying in the radial direction. The X axis coincides with the hexagonal axis of the *hcp* Co. The simulation shows that under such conditions the relaxed magnetization configuration of the cylinder depends on the K_{ME}, Figure 17. For $K_{ME} = 0$, the magnetization of the Co nanocylinder is homogeneous and oriented along the Z axis, Figure

17a. The presence of the magnetoelastic stresses, $K_{ME} = 5.35 \times 10^5$ J/m^3, breaks nanocylinder on two domains, Figure 17b.

The performed simulations of the magnetic structure of Co nanocylinders should be considered only as evaluation results showing that with the magnetoelastic component nanocylinders may be partitioned into domains, which leads to a decrease in remanence and coercivity and to isotropy of the magnetic properties. This is in good agreement with the experimental data.

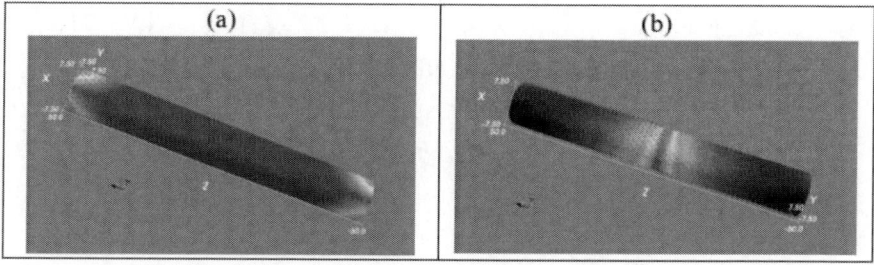

Figure 17. Relaxed magnetization configuration of *hcp* Co nanocylinder of diameter 20 nm and length 100 nm for the absence of stress, $K_{ME} = 0$ (a) and $K_{ME} = 5.35 \times 10^5$ J/m^3(b). Red color corresponds to the magnetization direction along the nanocylinder axis, blue color corresponds to the magnetization direction oriented perpendicular to the nanocylinder axis.

CONCLUSION

In this chapter we have presented different magnetic characteristics of CNT-based magnetic nanocomposites grown by CVD. Carbon nanotubes were vertically aligned. The arrangement of single domain ferromagnetic nanoparticles varied from one nanoparticle in each nanotube to several spaced nanoparticles localized within inner CNT channel and, finally, to an arbitrary distribution of nanoparticles throughout the sample array. The in-plane packing density of NPs was determined by the density of CNTs and was equal to appx. 10^{10}-10^{11} cm^{-2}. In all these samples magnetic properties were investigated paying special attention to the impact of the magnetic anisotropy. It was obtained that,

- for samples with one NP in each CNT the contribution of the magnetic anisotropy suppresses the magnetostatic dipole interaction between closely packed ferromagnetic NPs leading to the magnetic isolation of NPs;
- for samples with few NPs inside inner CNT channel the coherent magnetic anisotropy is of great importance causing microscopically large coherence of the local magnetic moments. These correlations could propagate up to micrometers. The RKKY interaction via the inner shell of CNTs could enforce this coherence;
- for samples with distributed NP`s arrangement through the whole sample the relaxation of the magnetic moments occurs via different metastable states. The magnetic anisotropy determines the temperature dependence of the magnetic viscosity in the intermediate temperature range.

ACKNOWLEDGMENTS

S. L. Prischepa and I. V. Komissarov acknowledge the financial support of the "Improving of the Competitiveness" Program of the National Research Nuclear University MEPhI – Moscow Engineering Physics Institute. The authors would like to thank F. Le Normand, J. Tejada and A. V. Kukharev for valuable discussions and suggestions.

REFERENCES

[1] Dorman, J. L., Fiorani, D. and Tronc, E. (1997). Magnetic relaxation in fine-particle systems. *Adv. Chem. Phys.*, 98: 283-494.
[2] Gubin, S. P., Koksharov, Y. A., Khomutov, G. B. and Yurkov, G. Yu. (2005). Magnetic nanoparticles: preparation, structure and properties. *Russ. Chem. Rev.*, 74 (6): 489-520.

[3] Proctor, T. C., Chudnovsky, E. M. and Garanin, D. A. (2015). Scaling of coercivity in a 3d random anisotropy model. *J. Magn. Magn. Mater.*, 384: 181-185.

[4] Liu, J., Qiao, S. Z., Hu, Q. H. and Lu, G. Q. (Max) (2011). Magnetic nanocomposites with mesoporous structures: synthesis and applications. *Small*, 7 (4): 425-443.

[5] Faupel, F., Zaporojtchenko, V., Strunkus, T. and Elbahri, M. (2010). Metal-polymer nanocomposites for functional applications. *Adv. Eng. Mater.*, 12 (12): 1177-1190.

[6] Wildgoose, J. J., Banks, C. E. and Compton, R. G. (2006). Metal nanoparticles and related materials supported on carbon nanotubes: methods and applications. *Small*, 2 (2): 182-193.

[7] Jourdain, V. and Bihara, C. (2013). Current understanding of the growth of carbon nanotubes in catalytic chemical vapor deposition. *Carbon*, 58: 2-39.

[8] Vorobjova, A. I., Shimanovich, D. L., Yanushkevich, K. I., Prischepa, S. L. and Outkina, E. A. (2016). Properties of Ni and Ni-Fe nanowires electrochemically deposited into a porous alumina template. *Beilstein J. Nanotechnol.*, 7: 1709-1717.

[9] Labunov, V. A., Danilyuk, A. L., Prudnikava, A. L., Komissarov, I. V., Shulitski, B. G., Speisser, C., Antoni, F., Le Normand, F. and Prischepa, S. L. (2012). Microwave absorption in nanocomposite material of magnetically functionalized carbon nanotubes. *J. Appl. Phys.*, 112 (2): 024302(1-9).

[10] Costa Jr., A. T., Kirwan, D. F. and Ferreira, M. S. (2005). Indirect exchange coupling between magnetic adatoms in carbon nanotubes. *Phys. Rev. B*, 72 (8): 085402(1-8).

[11] He, J., Zhou, P., Jiao, N., Ma, S. Y., Zhang, K. W., Wang, R. Z. and Sun, L. Z. (2014). Magnetic exchange coupling and anisotropy of 3d transition metal nanowires on graphyne. *Sci. Rep.*, 4: 4014(1-9).

[12] Costa, A. T., Rocha, C. G. and Ferreira, M. S. (2007). Noncollinear coupling between magnetic adatoms in carbon nanotubes. *Phys. Rev. B*, 76 (8): 085401(1-5).

[13] Klinovaja, J. and Loss, D. (2013). RKKY interaction in carbon nanotubes and graphene nanoribbons. *Phys. Rev. B*, 87 (4): 045422(1-11).
[14] Danilyuk, A. L., Prudnikava, A. L., Komissarov, I. V., Yanushkevich, K. I., Derory, A., Le Normand, F., Labunov, V. A. and Prischepa, S. L. (2014). Interplay between exchange interaction and magnetic anisotropy for iron based nanoparticles in aligned carbon nanotube arrays. *Carbon*, 68: 337-345.
[15] Prischepa, S. L., Danilyuk, A. L., Prudnikava, A. L., Komissarov, I. V., Labunov, V. A. and Le Normand, F. (2014). Exchange coupling and magnetic anisotropy for different concentration of iron based nanoparticles in aligned carbon nanotube arrays. *Phys. Status Solidi C*, 99 (5-6): 1074-1079.
[16] Chudnovsky, Eugene M. (1995). In: *The magnetism of amorphous metals and alloys*, edited by Fernandez-Baca, J. A. and Ching, W.-Y. (World Scientific, Singapore) Ch. 3: 143-175.
[17] Komogortsev, S. V., Iskhakov, R. S., Balaev, A. D., Kudashov, A. G., Okotrub, A. V. and Smirnov, S. I. (2007). Magnetic properties of Fe_3C ferromagnetic nanoparticles encapsulated in carbon nanotubes. *Phys. Sol. State*, 49 (4): 734-738.
[18] Varvaro, G., Albertini, F., Agostinelli, E., Casoli, F., Fiorani, D., Laureti, S., Lupo, P., Ranzieri, P., Astinchap, B. and Testa, A. M. (2012). Magnetization reversal mechanism in perpendicular exchanged-coupled Fe/L_{10}-FePt bilayers. *New J. Phys.* 14 (7): 073008(1-14).
[19] Navas, D., Torrejon, J., Béron, F., Redondo, C., Batallan, F., Toperverg, B. P., Devishvili, A., Sierra, B., Castaño, F., Pirota, K. R. and Ross C. A. (2012). Magnetization reversal and exchange bias effects in hard/soft ferromagnetic bilayers with orthogonal anisotropies. *New J. Phys.* 14 (11): 113001(1-21).
[20] Akulov, N. S. (1931). Zur Theorie der Magnetisierungskurve von Einkristallen. *Zeitschr. Physik.* 67 (11-12): 794-807.
[21] Brown Jr., W. T. (1940). Theory of the approach to magnetic saturation. *Phys. Rev.*, 58 (8): 736-743.

[22] Fähnle, M. and Kronmüller, H. (1978). The influence of spatially random magnetostatic, magnetocrystalline, magnetostrictive and exchange fluctuations on the law of approach to ferromagnetic saturation of amorphous ferromagnets. *J. Magn. Magn. Mater.*, 8 (2): 149-156.

[23] Chudnovsky, E. M. and Serota, R. A. (1983). Phenomenological theory of amorphous magnets with small random anisotropy. *J. Phys. C*, 16 (21): 4181-4190.

[24] Chudnovsky, E. M. (1989). Dependence of the magnetization law on structural disorder in amorphous ferromagnets. *J. Magn. Magn. Mater.*, 79 (1): 127-130.

[25] Martínez-Huerta, J. M., de la Torre Medina, J., Piraux, L. and Encinas, A. (2013). Configuration dependent demagnetizing field in assemblies of interacting magnetic particles. *J. Phys.: Condens. Matter*, 25 (22): 226003.

[26] Hiroi, K., Kura, H., Ogawa, T., Takahashi, M. and Sato, T. (2014). Magnetic ordered states induced by interparticle magnetostatic interaction in α-Fe/Au mixed nanoparticle assembly. *J. Phys.: Condens. Matter*, 26 (17): 176001.

[27] Son, Y.-W., Cohen, M. L. and Louie, S. G. (2006) Half-metallic graphene nanoribbons. *Nature (London)*, 444: 347-349.

[28] Lehtinen, P. O., Foster, A. S., Ayuela, A., Krasheninnikov, A., Nordlung, K. and Nieminen, R. M. (2003) Magnetic properties and diffusion adatoms on a graphene sheet. *Phys. Rev. Lett.*, 91 (1): 017202 (1-4).

[29] Giesbers, A. J. M., Uhlirová, K., Konečny, M., Peters, E. C., Burghard, M., Aarts, J. and Flipse, C. F. J. (2013). Interface-induced room-temperature ferromagnetism in hydrogenated epitaxial graphene. *Phys. Rev. Lett.*, 111 (16): 166101(1-5).

[30] Castro, E. V., Peres, N. M. R., Stauber, T. and Silva, N. A. P. (2008). Low-density ferromagnetism in biased bilayer graphene. *Phys. Rev. Lett.*, 100 (18): 186803(1-4).

[31] Pisani L., Montanari, B. and Harrison, N. M. (2008). A defected graphene phase predicted to be a room temperature ferromagnetic semiconductor. *New J. Phys.*, 10 (3): 033002(1-9).
[32] Danilyuk, A. L., Komissarov, I. V., Labunov, V. A., Le Normand, F., Derory, A., Hernandez, J. M., Tejada, J. and Prischepa, S. L. (2015). Manifestation of coherent magnetic anisotropy in a carbon nanotube matrix with low ferromagnetic nanoparticle content. *New J. Phys.*, 17 (2): 023073(1-12).
[33] Chudnovsky, E. M., Saslow, W. M. and Serota, R. A. (1986). Ordering in ferromagnets with random anisotropy. *Phys. Rev. B*, 33 (1): 251-261.
[34] Erdélyi, A. (Editor), (1954). *Tables of Integral Transforms*, 2, (McGraw-Hill Book Co, New York).
[35] Danilyuk, A. L., Komissarov, I. V., Kukharev, A. V., Le Normand, F., Hernandez, J. M., Tejada, J. and Prischepa, S. L. (2017). Impact of CNT medium on the interaction between ferromagnetic nanoparticles. *Europhys. Lett.*, 117: 27007(1-7).
[36] Prischepa, S. L., Danilyuk, A. L., Kukharev, A. V., Le Normand, F. and Cojocaru, C. S. (2019). Self-assembled magnetically isolated Co nanoparticles embedded inside carbon nanotubes. *IEEE Trans. Magn.*, 55: 2300304(1-4).
[37] Cojocaru, C. S. and Le Normand, F. (2006). On the role of activation mode in the plasma- and hot filaments-enhanced catalytic chemical vapour deposition of vertically aligned carbon nanotubes. *Thin Solid Films*, 515: 53-58.
[38] Mane, J. M., Cojocaru, C. S., Barbier, A., Deville, J. P., Thiodjio Sendja, B. and Le Normand, F. (2007). GISAXS study of the alignment of oriented carbon nanotubes grown on plain SiO_2/Si(100) substrates by a catalytically enhanced CVD process. *Phys. Status Solidi A*, 204 (12): 4209-4229.
[39] Le Normand, F., Cojocaru, C. S., Fleaca, C., Li, J. Q., Vincent, P., Pirio, G., Gangloff, L., Nedellec, Y. and Legagneux, P. (2007). A comparative study of the field emission properties of aligned films of

carbon nanostructures, from carbon nanotubes to diamond. *Eur. Phys. J. Appl. Phys.*, 38 (2): 115–127.

[40] Danilyuk, A. L., Kukharev, A. V., Cojocaru, C. S., Le Normand, F. and Prischepa, S. L. (2018). Impact of aligned carbon nanotubes array on the magnetostatic isolation of closely packed ferromagnetic nanoparticles. *Carbon*, 139: 1104-1116.

[41] Chudnovsky, E. M. (1988). Magnetic properties of amorphous ferromagnets. *J. Appl. Phys.*, 64 (10): 5770-5775.

[42] Harris, R., Plischke, M. and Zuckermann, M. J. (1973). New model for amorphous magnetism. *Phys. Rev. Lett.*, 31 (3): 160-162.

[43] Alben, R., Becker, J. J. and Chi, M. C. (1978). Random anisotropy in amorphous ferromagnets. *J. Appl. Phys.*, 49 (3):1653-1658.

[44] Iskhakov, R. S., Komogortsev, S. V., Balaev, A. D. and Chekanova, L. A. (2000). Dimensionality of a system of exchange-coupled grains and magnetic properties of nanocrystalline and amorphous ferromagnets. *JETP Letters*, 72 (6): 304-307.

[45] Ignatchenko, V. A., Iskhakov, R. S., Popov, G. V. (1982). Law of approach of magnetization to saturation in amorphous ferromagnets. *Sov. Phys. JETP*, 55 (5): 879-886.

[46] Tejada, J., Martinez, B., Labarta, A., Grössinger, R., Sassik, H., Vazguez, M. and Hernando, A. (1990). Phenomenological study of the amorphous $Fe_{80}B_{20}$ ferromagnet with small random anisotropy. *Phys Rev B*, 42 (1): 898-905.

[47] Chudnovsky, E. M. and Tejada, J. (1993). Evidence of the extended orientational order in amorphous alloys obtained from magnetic measurements. *Europhys. Lett.*, 23 (7): 517-522.

[48] Löffler, J. F., Meier, J. P., Doudin, B., Ansermet, J.-P. and Wagner, W. (1998). Random and exchange anisotropy in consolidated nanostructures Fe and Ni: role of grain size and trace oxides on the magnetic properties. *Phys Rev B*, 57 (5): 2915-2924.

[49] Prischepa, S. L. and Danilyuk, A. L. (2019). Anisotropic temperature-dependent interaction of ferromagnetic nanoparticles embedded inside CNT. *Int. J. Nanosci.*, 18 (3&4): 1940015(1-4).

[50] Klinovaja, J., Schmidt, M. J., Braunecker, B. and Loss, D. (2011). Helical modes in carbon nanotubes generated by strong electric fields. *Phys. Rev. Lett.*, 106 (15): 156809(1-4).

[51] Min, H., Hill, J. E., Sinitsyn, N. A., Sahu, B. B., Kleinman, L. and MacDonald, A. H. (2006). Intrinsic and Rashba spin-orbit interactions in graphene sheets. *Phys. Rev. B*, 74 (16): 165310(1-5).

[52] Zhou, W., Vavro, J., Nemes, N. M., Fischer, J. E., Borondicks, F., Kamarás, K. and Tanner, D. B. (2005). Charge transfer and Fermi level shift in p-doped single-walled carbon nanotubes. *Phys. Rev. B*, 71 (20): 205423(1-7).

[53] Kim, S., Jo, I., Dillen, D. S., Ferre, D. A., Fallahazad, B., Yao, Z., Banerjee, S. K. and Tutuc, E. (2012). Direct measurement of the Fermi energy in graphene using a double-layer heterostructure. *Phys. Rev. Lett.*, 108 (11): 116404(1-4).

[54] Steele, G. A., Pei, F., Laird, E. A., Jopi, J. M., Meerwaldt, H. B. and Kouwenhoven, L. P. (2013). Large spin-orbit coupling in carbon nanotubes. *Nat. Commun.*, 4 (1): 1573.

[55] Collins, P. G., Arnold, M. S. and Avouris, P. (2001). Engineering carbon nanotubes and nanotube circuits using electrical breakdown. *Science*, 292 (5517): 706-709.

[56] Stetter, A., Vancea, J. and Back, C. H. (2008). Determination of the intershell conductance in a multiwall carbon nanotube. *Appl. Phys. Lett.*, 93 (17): 172103(1-3).

[57] Prudnikava, A. L., Fedotova, J. A., Kasiuk, J. V., Shulitski, B. G. and Labunov, V. A. (2010). Mössbauer spectroscopy investigation of magnetic nanoparticles incorporated into carbon nanotubes obtained by the injection CVD method. *Semiconductor Physics, Quantum Electronics & Optoelectronics*, 13 (2): 125-131.

[58] Ferrari, A. C. and Robertson, J. (2000). Interpretation of Raman spectra of disordered and amorphous carbon. *Phys. Rev. B*, 61 (20): 14095-14107.

[59] Pimenta, M. A., Dresselhaus, G., Dresselhaus, M. S., Cancado, L. G., Jorio, A. and Saito, R. (2007). Studying disorder in graphite-based

systems by Raman spectroscopy. *Phys. Chem. Chem. Phys.*, 9 (11): 1276-1291.

[60] Di Leo, R. A., Landi, B. J. and Raffaelle, R. P. (2007). Purity assessment of multiwalled carbon nanotubes by Raman spectroscopy. *J. Appl. Phys.* 101 (6): 064307.

[61] Pawlyta, M., Rouzaud, J. N. and Duber, S. (2015). Raman microspectroscopy characterization of carbon black: Spectral analysis and structural information. *Carbon*, 84: 479–490.

[62] Saito, A., Hofmann, M., Dresselhaus, G., Jorio, A. and Dresselhaus, M. S. (2011). Raman scattering of graphene and carbon nanotubes. *Adv. Phys.*, 60 (3): 413–550.

[63] Bokobza, L., Bruneel, J.-L. and Michel Couzi, M. and Raman, C. (2015). Spectra of Carbon-Based Materials (from Graphite to Carbon Black) and of Some Silicone Composites. *J. Carbon Research*, 1 (1): 77-94.

[64] Byeon, J. W. and Kwun, S. I. (2003). Magnetic evaluation of microstructures and strength of eutectoid steel. *Materials Transactions*, 44 (10): 2184-2190.

[65] Pfeiffer, H. and Schüppel, W. (1990). Investigation of magnetic properties of barium ferrite powders by remanence curves. *Phys. Status Solidi A*, 119 (1): 259-269.

[66] Tejada, J. and Zhang, X. (1995). Experiments in quantum magnetic relaxation. *J. Magn. Magn. Mater.*, 140-144 (Part 3): 1815-1818.

[67] Tejada, J. and Chudnovsky, E. M. (1998). *Macroscopic Quantum Tunneling of the Magnetic Moment*. Cambridge University Press.

[68] Tejada, J., Zhang, X. X. and Chudnovsky, E. M. (1993). Quantum relaxation in random magnets. *Phys. Rev. B*, 47 (22): 14977-14987.

[69] Grobert, N., Hsu, W. K., Zhu, Y. Q., Hare, J. P., Kroto, H. W. and Walton, D. R. M. (1999). Enhanced magnetic coercivities in Fe nanowires. *Appl. Phys. Lett.*, 75 (21): 3363-3365.

[70] Leonhardt, A., Ritschel, M., Kozhuharova, R., Graff, A., Mühl, T., Huhle, R., Mönch, I., Elefant, D. and Schneider, S. M. (2003). Synthesis and properties of filled carbon nanotubes. *Diam. Rel. Mater.*, 12 (3-7): 790-793.

[71] Baaziz, W., Begin-Colin, S., Pichon, B. P., Florea, I., Ersen, O., Zafeiratos, S., Barbosa, R., Begin, D. and Pham-Huu, C. (2012). High-density monodispersed cobalt nanoparticles filled into multiwalled carbon nanotubes. *Chem. Mater.*, 24 (9): 1549-1551.

[72] Ghunaim, R., Eckert, V., Scholz, M., Gellesch, M., Wurmehl, S., Damm, C., Büchner, B., Mertig, M. and Hampel, S. (2018). Carbon nanotube-assisted synthesis of ferromagnetic Heusler nanoparticles of Fe_3Ga (Nano-Gajfenol). *J. Mater. Chem. C*, 6 (5): 1255-1263.

[73] Vázquez, M., Pirota, K., Torrejón, J., Navas, D. and Hernández-Vélez, M. (2005). Magnetic behavior of densely packed hexagonal arrays of Ni nanowires: influence of geometric characteristics. *J. Magn. Magn. Mater.*, 294 (2): 174-181.

[74] Li, C., Wu, Q., Yue, M., Xu, H., Palaka, S., Elkins, K. and Liu, J. P. (2017). Manipulation of morphology and magnetic properties in cobalt nanowires. *AIP Advances*, 7 (5): 056229(1-5).

[75] Girt, E., Huttema, W., Mryasov, O. N., Montoya, E., Kardasz, B., Eyrich, C., Heinrich, B., Dobin, A. Yu. and Karis, O. (2011). A method for measuring exchange stiffness in ferromagnetic films. *J. Appl. Phys.*, 109 (7): 07B765(1-3).

[76] Zaluska-Kotur, M. A. and Cieplak M. (1993). Dipole interaction with random anisotropy - a local-mean-field study. *Europhys. Lett.*, 23 (2): 85-90.

[77] Russier, V., Petit, C., Legrand, J. and Pileni, M. P. (2000). Collective magnetic properties of cobalt nanocrystals self-assembled in a hexagonal network: Theoretical model supported by experiments. *Phys. Rev. B*, 62 (6): 3910-3916.

[78] Kechrakos D. and Trohidou, K. N. (2008). Dipolar interaction effects in the magnetic and magnetotransport properties of ordered nanoparticle arrays. *J. Nanosci. Nanotechnol.*, 8 (6): 2929-2943.

[79] Vidal, F., Zheng, Y., Schio, P., Bonilla, F. J., Barturen, M., Milano, J., Demaille, D., Fonda, E., de Oliveira, A. J. A. and Etgens, V. H. (2012). Mechanism of localization of the magnetization reversal in 3 nm wide Co nanowires. *Phys. Rev. Lett.*, 109 (11): 117205(1-5).

[80] Sánchez-Barriga, J., Lucas, M., Radu, F. Martin, E., Multigner, M., Marin, P., Hernando, A. and Rivero, G. (2009). Interplay between the magnetic anisotropy contributions of cobalt nanowires. *Phys Rev B*, 80 (18): 184424(1-8).

[81] Sander, D. (1999). The correlation between mechanical stress and magnetic anisotropy in ultrathin films. *Rep. Prog. Phys.*, 62 (5): 809-858.

[82] Lee, C. H., He, H., Lamelas, F. J., Vavra, W., Uher, C. and Clarke, R. (1990). Magnetic anisotropy in epitaxial Co superlattices. *Phys. Rev. B*, 42 (1): 1066-1069.

[83] Yu, M. F., Lourie, O., Dyer, M. J., Moloni, K., Kelly, T. F. and Ruoff, R. S. (2000). Strength and breaking mechanism of multi-walled carbon nanotubes under tensile load. *Science*, 287 (5453): 637–640.

[84] http://nmag.soton.ac.uk/nmag/.

In: A Closer Look at Magnetic Anisotropy　　ISBN: 978-1-53617-566-0
Editor: Georges Fremont　　© 2020 Nova Science Publishers, Inc.

Chapter 2

ENHANCEMENT OF PERPENDICULAR MAGNETIC ANISOTROPY IN FECOZR-CAF$_2$ NANOCOMPOSITE FILMS BY COMBINED INFLUENCE OF NANOPARTICLES OXIDATION AND ION IRRADIATION

Julia Kasiuk[1], Julia Fedotova[1,], Vadim Bayev[1], Janusz Przewoźnik[2], Czesław Kapusta[2], Vladimir Skuratov[3], Momir Milosavlievič[4] and Jacques O'Connell[5]*

[1]Institute for Nuclear Problems of Belarusian State University, Minsk, Belarus
[2]AGH University of Science and Technology, Kraków, Poland
[3]Joint Institute for Nuclear Research, Dubna, Russia
[4]VINČA Institute of Nuclear Sciences, Belgrade University, Belgrade, Serbia
[5]N. Mandela Metropolitan University, Port Elizabeth, South Africa

---

[*] Corresponding Author's E-mail: julia@hep.by.

ABSTRACT

The study presented is focused on the aspects of perpendicular magnetic anisotropy in FeCoZr-CaF$_2$ nanocomposite films induced by shape anisotropy of metallic nanoparticles. It also concers the methods of anisotropy enhancement through the films treatment by means of oxidation of metallic nanoparticles and heavy ions irradiation of nanocomposites, as well as the combination of both procedures. Oxidation and irradiation conditions, i.e., the oxygen pressure P_O during films sputtering and Xe ions fluence D, are under consideration for maximal magnetic anisotropy increase. The influence of oxidation and heavy ions bombardment on the crystalline structure, phase composition and magnetic state of nanoparticles as well as on magnetic parameters of the composite films (in particular, the anisotropy and demagnetizing fields) characterizing their anisotropic properties are studied by transmission electron microscopy, X-ray diffraction, Mössbauer spectroscopy and magnetometry. A considerable enhancement of the perpendicular magnetic anisotropy is found in the case of "ferromagnetic α-FeCo(Zr) core − antiferromagnetic α-Fe$_2$O$_3$ shell" structure of nanoparticles formed at P_O = 4.3 mPa. It corresponds to more than two-fold increase in the films anisotropy field (up to 3.5 kOe), as compared with that containing non-oxidised ferromagnetic nanoparticles and to a decrease in the the canting angle of their magnetic moments (from 24 to 16°) with respect to the direction of the film normal. Nonmagnetic channels formation is believed to be responsible for the magnetic nanoparticles separation and for the corresponding decrease in the films demagnetizing field as the result of their irradiation by Xe ions with moderate fluences (less than $1 \cdot 10^{13}$ ion/cm^2) that provides increase of perpendicular component of magnetic anisotropy. Protective function of the oxide shells is revealed in the case of high Xe fluence application (up to $2.5 \cdot 10^{13}$ ion/cm^2) that prevents destruction of ordered nanoparticles. Improvement of nanoparticles crystallinity is found in the case of irradiation of the films with complex "core-shell" structure of nanoparticles and is accompanied by the preservation of high perpendicular magnetic anisotropy.

Keywords: perpendicular magnetic anisotropy, nanocomposites, core-shell structure, oxidation, ion irradiation

INTRODUCTION

Perpendicular magnetic anisotropy (PMA) media are of practical interest for high density recording and sensing applications. Search for new types of anisotropic materials is one of the main tasks for PMA enhancement and decrease in magnetic structures (domains) size. Nanocomposite metal-insulator materials exhibiting shape anisotropy of magnetic nanoparticles (NPs) are promising candidates for PMA media because of extremely low nanostructures sizes of a few nanometers and reasonable values of anisotropy fields H_a induced by elongated NPs shape. Overcoming the problem of superparamagnetic limit is associated with NPs magnetic moments stabilization by the anisotropy field in the direction of NPs long axes. Crucial parameter characterizing PMA of such systems is the aspect ratio of NPs axes describing their elongation. Detailed and systematic investigation of correlation between PMA and structure of granular metal-insulator films reported in [1, 2] proves that formation of NPs with columnar-like shape is governed by the composition of magnetic NPs and non-magnetic matrix, as well as by their relative concentration. Namely, NPs shape originates from self-organization of different type atoms during films nucleation and further growth. In particular, $(FeCoZr)_x(CaF_2)_{100-x}$ films demonstrate PMA provided by the elongation of α-FeCo(Zr) nanoparticles in the range of x = 58-73 at.%, whereas the films with $x \leq 39$ at.% contain spherical particles [1]. Evolution of NPs shape with x in $(FeCoZr)_x(CaF_2)_{100-x}$ films is shown schematically in Figure 1 and is supported by cross-sectional images obtained by transmission electron microscopy (TEM) for x = 39 and x = 73 at.% films.

Additionally, NPs elongation with increasing x is proved by magnetic resonance spectroscopy [3] providing data on NPs aspect ratio a/c, as well as by magnetometry detecting increase in H_a [1]. One more important parameter influencing on PMA is the distance d between magnetic nanostructures. Unfortunately, elongation of NPs with x is accompanied by decrease in d that leads to enhancement of magnetostatic interaction between NPs and corresponding increase in demagnetizing field H_d in the films. The latter provides undesirable decrease in PMA.

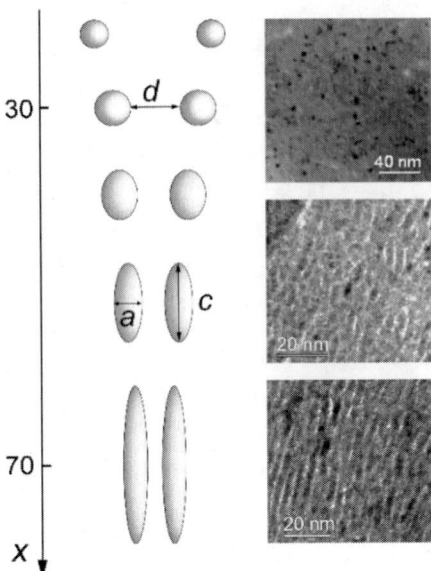

Figure 1. Schematic illustration of NPs shape evolution with x in $(FeCoZr)_x(CaF_2)_{100-x}$ nanocomposite films; the insets present cross-sectional TEM images of $x = 39$, $x = 58$ and $x = 73$ films.

One more problem concerning nanocomposites with PMA is caused by significant irregularities in the distribution of magnetic nanostructures as well as by inhomogenities in grain orientations due to natural randomness. Angular deviations of NPs long axes from the direction normal to the film plane considerably decrease PMA. Basing on TEM and Mössbauer spectroscopy results [1, 3], we have found that mean deviation of axes in $(FeCoZr)_x(CaF_2)_{100-x}$ films varies in the range of angles α of 5-30° for x = 58-73 at.% compositions. Thus, the methods of nanocomposites treatment that allow a decrease in such deviations are of critical importance.

One of the methods commonly used for NPs sizes homogenization is their partial oxidation leading to "core-shell" structures formation [4, 5]. The idea of application of oxidation to nanocomposies with PMA stems from the need for separation of elongated nanostructures from each other by non-ferromagnetic oxide shells that serve as additional barriers for NPs magnetic interaction. Such a separation is schematically presented in Figure 2a.

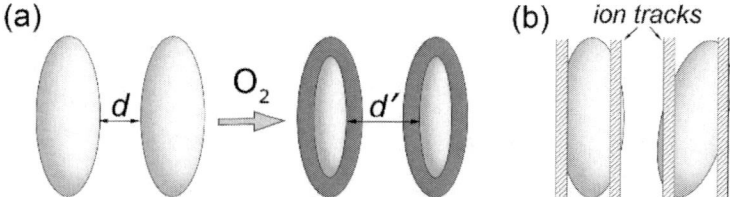

Figure 2. Schematic illustration of partial oxidation (a) and irradiation (b) for intended nanoparticles elongation.

Another method for inducing PMA proposed in this work is based on the irradiation of the films by heavy ions which are supposed to generate non-magnetic tracks [3, 6, 7]. Similarly to the oxide shells, their main function is in separation of NPs, and additionally, in NPs "drilling" in the direction of ion beam, i.e., along the films normal. As presented in Figure 2b, such "drilling" is believed to provide the aligning NPs along the films normal direction. Tuning the beam for appropriate density and diameters of ion tracks is supposed to induce PMA enhancement in the films studied.

This chapter is devoted to the analysis of the results of FeCoZr-CaF$_2$ films treatment by means of both above mentioned methods as well as their combination intended for PMA enhancement. Optimization of treatment conditions, namely oxygen pressure during films sputtering and ions fluence at subsequent irradiation, is worked out in details for obtaining increased PMA and appropriate magnetic parameters of the films studied.

EXPERIMENTAL DETAILS

The $(FeCoZr)_x(CaF_2)_{100-x}$ nanocomposite films ($29 \leq x \leq 73$ at. %) of the thickness 3-6 μm were deposited by DC ion sputtering, using 2 keV argon ion gun, at 0.28 nm/s, onto Al uncooled substrates (T_{sub} ~373 K) according the procedure described in [1]. The Ar pressure in the chamber was 67 mPa. Composition of $(FeCoZr)_x(CaF_2)_{100-x}$ films was verified by energy dispersive X-ray spectroscopy (EDX) in a scanning electron microscope. The films with $x = 58$ and 73 at.% compositions were chosen in the present study for detailed analysis because of maximal PMA detected for them in the previous works [1, 8]. Hereafter, the films of this compositions will be marked briefly as $x = 58$ film and $x = 73$ film. Irradiation of $(FeCoZr)_x(CaF_2)_{100-x}$ nanocomposite films ($x = 58, 73$ at.%) was carried out by 167 MeV Xe^{26+} ions with fluence D in the range of $5 \cdot 10^{12}$–$2.5 \cdot 10^{13}$ ion/cm^2 generated by IC-100 heavy ion cyclic accelerator (JINR, Dubna). Orientation of ion beam is always along the normal of the films. For NPs oxidation, oxygen was added in sputtering atmosphere with the partial pressure of $P_O = 4,3$ mPa. As it was previously shown [4, 8], deposition of $(FeCoZr)_x(CaF_2)_{100-x}$ films in such ambient leads to partial NPs oxidation with ferromagnetic properties conservation, whereas higher oxygen pressure (9,8 mPa) occurs in full NPs oxidation accompanied by their ferromagnetic properties transformation to paramagnetic (or antiferromagnetic) [4].

X-ray diffraction (XRD) analysis was done with an Empyrean PANalytical diffractometer using a diffracted beam graphite monochromator and an X'Celerator linear detector (Cu K$_\alpha$ radiation). The data were collected with divergent-beam optics (using 1/32" divergence and 1/16" anti-scatter slits) at a grazing incidence of 5 degrees with respect to the sample surface, with the detector scanning the 2Θ space over the 10-120 degree range. Lanthanum hexaboride (LaB$_6$ – NIST, Standard Reference Material 660a) was used to determine instrumental broadening of diffraction peaks. The experimental data were analysed using the profile fitting program *FullProf* [9] based on the Rietveld method. The background intensity was refined with a polynomial and the peak shape was approximated with a pseudo-Voigt function.

Microstructural studies were carried out by transmission electron microscopy (TEM) at normal and high resolution (HRTEM), using Philips EM400T microscope operated at 120 kV as well as Philips CM200 and JEOL ARM200F microscopes operated at 200 kV. The films were prepared by ion-beam thinning for cross-sectional view. For analysis of NPs chemical composition, electron energy loss spectroscopy (EELS) was carried out using a GATAN GIF Quantum 965ERS with dual EELS capability.

The magnetic moments (μ) were measured by DC magnetization measurements in the temperature range from 2 K up to 350 K and in magnetic fields H up to 90 kOe using the Vibrating Sample Magnetometer (VSM) option of the Quantum Design Physical Property Measurement System (PPMS). During measurements, magnetic field H was applied in parallel ($H_{\parallel}$) and perpendicular ($H_{\perp}$) directions to the thin film surface.

The ^{57}Fe transmission Mössbauer spectroscopy measurements were performed using a conventional constant acceleration type spectrometer with a 20 mCi ^{57}Co in Rh source moving at room temperature (RT). The absorber was fixed in a top loading type cryostat and spectra were recorded at RT and at 78 K. The spectra were analysed with MOSMOD software assuming distributions of effective hyperfine magnetic field (B_{hf}) and quadrupole splitting ($\Delta=(eQV_{zz}/2)(1+\eta^2/3)^{1/2}$, where e is elemental charge, Q is quadrupole moment of ^{57}Fe, V_{zz} is the maximal value of electric field gradient, η is the asymmetry parameter) based on the method published in [10]. Values of center shift δ are given with respect to α-Fe at RT.

RESULTS AND DISCUSSION

Structure and PMA of FeCoZr-CaF$_2$ Nanocomposites

Bright field cross-section TEM images of (FeCoZr)$_x$(CaF$_2$)$_{100-x}$ nanocomposite films with $x = 58$ (a) and 73 (b) are presented in Figure 3a and b, respectively. Top panels of Figure 3 demonstrate numerous metallic nanostructures of dark gray color characterizing microstructure typical for the whole films studied. NPs of elongated columnar-like shape are oriented almost normally to the films substrates [1]. More detailed HRTEM images of the films are shown in the middle panel of Figure 3.

As can be seen from HRTEM images, there is no substantial difference between microstructure of the films of both compositions. In both cases, 3-4 nm in cross-section metallic NPs are separated by non-metallic matrix and extended through the whole thickness of the films. However, the density of magnetic nanostructures is slightly different for the films studied, i.e., the distances between nanocolumns are smaller in the case of $x = 73$ film. Additionally, it should be noted that metallic nanocolumns presented in Figure 3 (middle panels) consist of the chains of discrete randomly oriented crystallites [1]. Phase identification for the studied films carried out by XRD (lower panels of Figure 3) confirms crystalline structure of *bcc* α-FeCo(Zr) nanoparticles and *fcc* CaF_2 matrix. Decrease in CaF_2 phase contribution with x is also obvious from XRD patterns. Relative intensities of diffraction lines characterizing α-FeCo(Zr) phase in Figure 3b indicates no preferential crystalline orientation, i.e., isotropic crystalline structure of NPs that confirms random orientation of crystallites in nanocolumns observed by HRTEM.

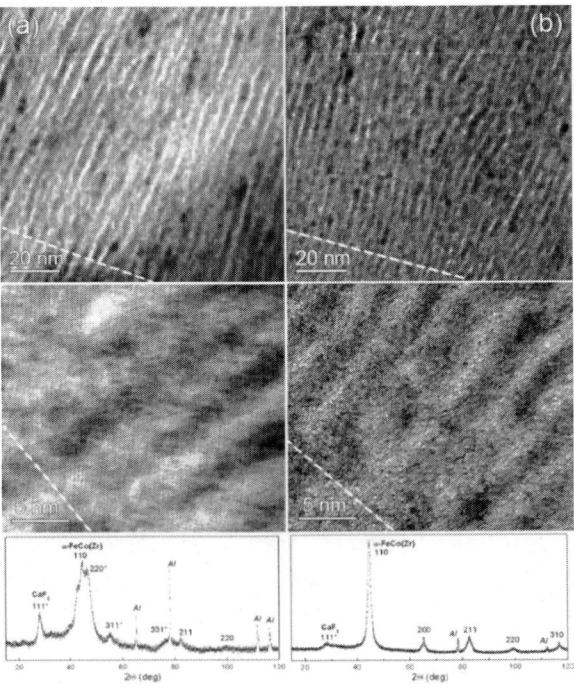

Figure 3. Cross-sectional TEM images (upper panels), HRTEM images (middle panels) and XRD patterns (lower panels) of $x = 58$ (a) and $x = 73$ (b) nanocomposite films. Dashed lines in TEM and HRTEM images indicates the orientation of Al substrate.

It is important to note, basing on TEM and HRTEM images, that the films contains numerous structural inhomogeneities and imperfections, such as local chaotic deviations of nanocolumns growth orientations from the films normal, discontinuities of metallic columns interrupted by CaF_2 matrix regions, columns cross-section size dispersions, etc. Such imperfections are supposed to decrease average magnetic anisotropy of the whole system induced by shape anisotropy of NPs.

The peculiarities of shape anisotropy of NPs in the films were additionally analyzed by Mössbauer and ferromagnetic-resonance (not shown, see [3]) spectroscopies. Mössbauer spectra of $x = 58$ and 73 films studied are presented in Figures 4, (a) and (b), respectively. They are measured without an external magnetic field in the geometry perpendicular to the films surface. As can be seen, both samples are mainly characterized by ferromagnetically split sextet corresponding to α-Fe-based phase with $\delta = 0.03$-$0.05(1)$ mm/s and $<B_{eff}> = 321$-$331(2)$ kGs, depending on x [1]. Additionally, small amount of superparamagnetic α-Fe-based phase (4%) described by doublet is detected in $x = 58$ film. Magnetic anisotropy of iron-containing nanostructures can be evaluated from the ratio of spectral lines intensities (by occupied square) in magnetically split sextets. Thus, $h_1:h_2:h_3 = 3:0.52:1$ and $h_1:h_2:h_3 = 3:0.37:1$ ratios (in arbitrary units) were derived from RT spectra of $x = 58$ and 73 films, respectively. K value equal to h_2/h_3 allows calculation the angle α determining preferential orientation of NPs magnetic moments with respect to the films normal averaged over the sphere basing on the relation $\alpha = \arccos\{[(4-K)/(4+K)]^{1/2}\}$ [11]. The values of α for $x = 58$ and 73 films amounts to 29 and 24°, respectively. One of the possible reasons of high α values characterizing deviation of NPs easy axes from the film normal is in the dispersion in the growth orientations of nanocolumns that was estimated from TEM images as 15-20° for both samples independently on x because of stochastic nature of such deviations [1]. It should be mentioned, that some difference in α value obtained by both methods is partly explained by the averaging procedure – (*i*) over the sphere in the case of Mössbauer spectroscopy and (*ii*) in plane for TEM images analysis.

One more possible interpretation of high α values obtained from Mössbauer spectra is in the deflection of preferential orientations of magnetic moments from the direction of nanocolumns growth. Indeed, ferromagnetic-resonance spectroscopy establishes substantial difference in the shape of magnetic NPs in the films studied [3]. Calculations based on its results show that the shape of averaged ferromagnetic nanoparticle in x = 58 film is simulated by elongated ellipsoid with aspect ratio ~1.2:1 and by infinite cylinder for x = 73 film [3]. These data correlates with the results of Mössbauer spectroscopy, because NPs with lower aspect ratio demonstrate weaker anisotropy. Moreover, in the chains of NPs with the shape close to spherical, crystalline anisotropy becomes more pronounced, the orientation of crystallites being random [1].

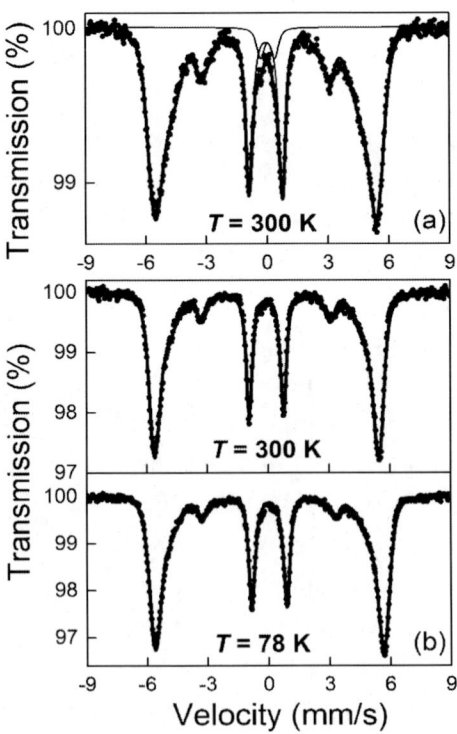

Figure 4. Mössbauer spectra of x = 58 (a) and x = 73 (b) films.

To clarify a possibility of thermal fluctuations influence on NPs magnetic moments orientation, low temperature Mössbauer spectrum ($T = 78$ K) of $x = 73$ film was analyzed (Figure 4b). K value of 0.30 derived from the sextet lines intensities leads to $\alpha = 22°$. Thus, decrease in temperature from 300 to 78 K results in 2° reduction of magnetic moments dispersion.

Magnetic field dependences of $x = 58$ and 73 films magnetic moments $\mu(H)$ are depicted in Figure 5. External magnetic field H is applied in the direction of the films plane $H_\parallel$ (Figure 5a) and along their normal $H_\perp$ (Figure 5b). If applied in the films plane, H is oriented mainly in perpendicular to nanocolumns long axes direction, i.e., in hard magnetization direction.

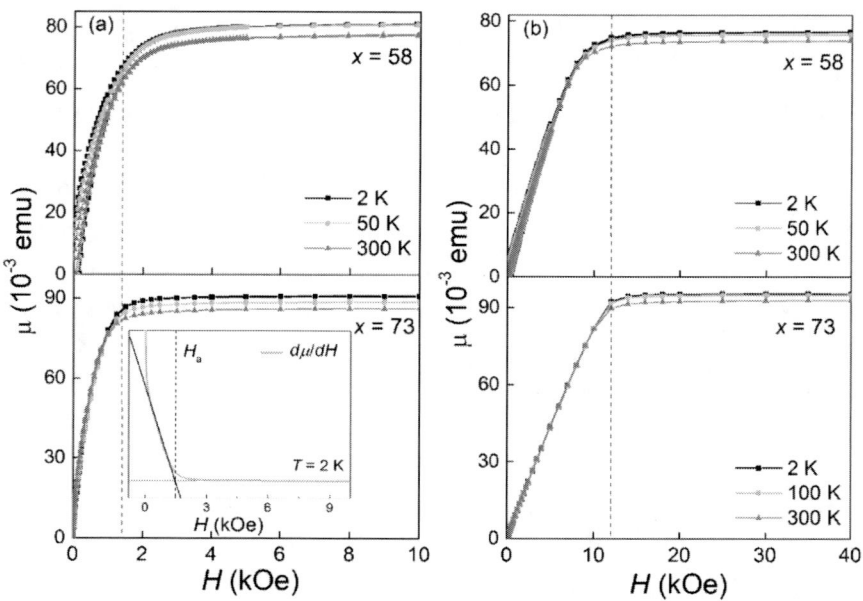

Figure 5. $\mu(H)$ dependences of $x = 58$ and $x = 73$ films measured in the films plane (a) and along the films normal (b) at different temperatures; dash lines correspond to H_a (a) and H_d (b) values of $x = 73$ film. The inset in (a) shows the first derivative of $\mu(H)$ curve of $x = 73$ film measured at $T = 2$ K for illustration of the procedure of H_a determination (similarly for H_d).

The most important parameter characterizing magnetic anisotropy of nanocolumns, namely, anisotropy field H_a can be determined directly from $\mu(H_{\parallel})$ dependences as the field magnetizing nanostructures up to saturation in hard magnetization direction. Since nanocolumns demonstrate some dispersion in growth orientations and, correspondingly, in their hard axes directions, there is no sharp inflection point in $\mu(H_{\parallel})$ dependences at saturation, especially in the case of $x = 58$ film. This means that H_a parameter can be estimated by this method with some inaccuracy, but more precisely for $x = 73$ film (dash line in Figure 5a). As can be seen from Figure 5a, H_a demonstrates higher value for $x = 58$ film, as compared with $x = 73$ sample, simultaneously accompanied by its higher distribution. The former is the result of higher distances between magnetic nanostructures in $x = 58$ film and their lower interaction that allows nanocolumns to demonstrate their individual properties in higher extent. High distribution of H_a is provided by the dispersion in the shape of magnetic NPs which are more spherical in $x = 58$ film according to ferromagnetic-resonance and Mössbauer spectroscopies data.

Dense "packing" of nanocolumns in the composite films leads to their intensive magnetostatic interaction. Such a system of magnetically interacting nanostructures responds to the external magnetic field as the whole, i.e., as the quasi-uniform ferromagnetic material. According to the Néel approach [12] which introduces average magnetostatic field setting by nanoparticle neighbors and influencing on it, H_a of assembly of nanocolumns can be determined as

$$H_a = 4\pi \cdot \Delta N \cdot M_s (1 - f_v) \tag{1}$$

where $\Delta N = N_{IP} - N_{OP}$ is the difference in demagnetizing factors of nanoparticle in two orthogonal orientations (IP – in the film plane and OP – along its normal), $M_S = 1330$ emu/cm^3 is saturation magnetization of α-FeCo(Zr) nanoparticles [1] and f_v is volume fraction of magnetic NPs. Using the results of ferromagnetic-resonance spectroscopy [3] providing $\Delta N = 0.5$ value and $f_v = 0.59$ calculated for $x = 73$ film, corresponding H_a

value amounts to 3.5 kOe. As can be seen from Figure 5a, estimated from $\mu(H_\parallel)$ curve H_a value for $x = 73$ film does not exceed 1.5 kOe. Such disagreement probably can be explained by more intensive magnetic interaction than Néel model assumes as well as by numerous structural imperfections evident from TEM images which are not under consideration in NPs modeling by infinite cylinder.

Magnetization curves $\mu(H_\perp)$ measured along the films normal (Figure 5b) also demonstrate high saturation field exceeding 11-12 kOe that is provided by demagnetizing effects. Rather low thickness of the films appears in high N_{OP}^{film} value close to 4π. Demagnetizing field H_d provided by the film shape and acting to the NPs magnetic moments when saturated along their long axes can be calculated according to $H_d = 4\pi \cdot M_S \cdot f_V$. It amounts to 9.7 kOe for $x = 73$ film ($f_v = 0.59$). Similarly to the case of H_a, discrepancy of calculated value and the parameter evaluated from $\mu(H_\perp)$ dependence ($H_d \sim 11$ kOe for $x = 73$ film) indicates close conglomeration of magnetic NPs and stronger interaction between them than it is expected for the film with $f_v = 0.59$.

To separate the effects of nanocolumns magnetic anisotropy and demagnetization provided by the film shape, the procedure of $\mu(H_\perp)$ dependences correction in the assumption of self-consistent fields [2, 13] was carried out. Indeed, demagnetizing field H_d acts to NPs in the direction opposite to their magnetic moments and decreases the influence of external magnetic field H. Thus, "internal" magnetic field H^i influencing on nanocolumns magnetic moments is determined as

$$H^i = H - 4\pi \cdot M(H) \cdot f_v, \qquad (2)$$

where magnetization M changes with H. The result of field correction for $x = 73$ film is presented in Figure 6a. The $\mu(H_\perp)$ dependences before and after correction are shown supplemented with $\mu(H_\parallel)$ curve for comparison.

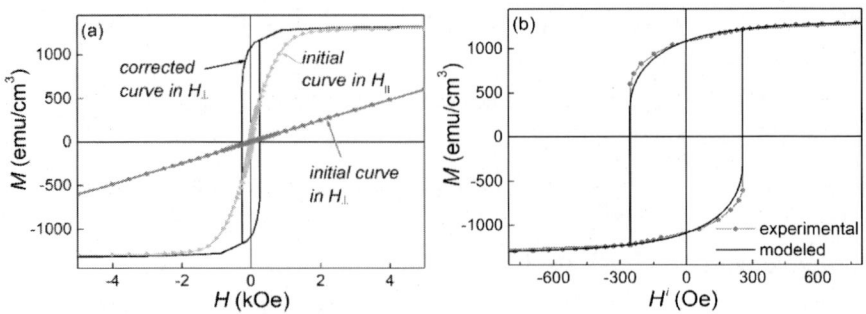

Figure 6. Parts of μ($H_\perp$) dependences of $x = 73$ film corrected for demagnetizing field effect (a) and modeled with Stoner-Wohlfarth dependence for prolate spheroid (b).

As can be observed in Figure 6a, correction leads to the significant changes in the magnetization curve shape. The μ($H_\perp^i$) dependence corresponds now to easy magnetization axis of nanocolumns, as compared with μ(H_{II}) curve. However, the shape of derived curve does not completely agree with typical easy magnetization curve obtained by Stoner and Wohlfarth [14] for ferromagnetic prolate spheroid. Squareness of the curve determined as M_r/M_S (M_r is remanent magnetization) is equal to 0.82. The M_r/M_S parameter less than 1 indicates that magnetization curve is measured in the orientation tilted from easy magnetization axis. Corrected μ($H_\perp^i$) curve can thus be fitted in the frame of Stoner-Wohlfarth (S&W) model for prolate spheroid, which is magnetized in the direction tilted from its long axis by α angle [14, 15]. In this case, its magnetization reversal is described by

$$h \cdot \sin\theta + \frac{1}{2}\sin 2(\theta - \alpha) = 0, \qquad (3)$$

where $h = H^i/4\pi \cdot \Delta N \cdot M_S$ is normalized magnetic field; θ is the angle between magnetic field h and nanocolumn magnetic moment; α is the angle between h and easy magnetization axis of nanocolumn (or prolate spheroid), i.e., parameter α gives the average orientation of nanocolumns magnetic moments with respect to the film normal, since h is oriented along this direction. The approximating curve is presented in Figure 6b.

Close fit confirms that magnetization reversal of nanocolumns is realized dominantly by means of coherent rotation mechanism described by Stoner-Wohlfarth model. Parameters of magnetic anisotropy H_a and α derived from the fit are evaluated as 500 Oe and 35°, respectively. It should be mentioned, that estimated parameters demonstrates some discrepancy with the results obtained by other methods. Overestimated α value compared to ~15° from TEM and 24° from Mössbauer spectroscopy, as well as understated H_a parameter with respect to 1.5 kOe corresponding to the inflection point of $\mu(H_{\textrm{ll}})$ curve (Figure 5a) is the consequence of α and H_a "modification" by magnetostatic NPs interaction. As it was previously establishyed by Wohlfarth [12], magnetization reversal of magnetically interacting elongated NPs can also be described by (3), but the angle α should be substituted for α+α', where α is the angle between h and easy magnetization axis of nanocolumn, whereas α' is an additional angle appearing in the result of NPs interaction, i.e., interaction "modifies" the balance orientation of nanoparticle magnetic moment and tilts it by the extra angle α'. Parameter h is also replaced by h' taking into account NPs interaction. Precise determination of H_a and α values free from interaction influence basing on Wohlfarth model [12] is complicated and depends on relative positions and orientations of NPs. It should be mention that α' becomes zero only in the case of strongly perpendicular orientation of nanocolumns to the film surface that is not realized in the present case. Estimation of α and H_a parameters is proposed to be derived from the equation of energy balance of the system using boundary conditions, such as zero field ($H = 0$) and magnetization saturation ($\theta = 0$). Magnetic energy of the system per unit volume in $H_\perp$ field can be expressed as

$$\varepsilon = K_u \cdot \sin^2(\theta - \alpha) + K_f \cdot \cos^2\theta - M_s \cdot H_\perp \cdot \cos\theta, \qquad (4)$$

where $K_u = 2\pi \cdot \Delta N \cdot M_S^2$ is the constant of uniaxial shape anisotropy, $K_f = 2\pi \cdot M_S^2 \cdot f_V^2$ is the constant characterizing magnetostatic interaction of NPs in the film, θ and α are, as was previously determined, the angles between the applied field $H_\perp$ and nanocolumns magnetic moments, as well as NPs shape anisotropy direction, respectively. In this approach, NPs interaction tilting their magnetic moments closer to the film plane is considered as the uniaxial anisotropy with the constant K_f and easy axis oriented in the film plane. The balance between shape anisotropy and magnetostatic energy in zero magnetic field can be described with [15]

$$\frac{\partial \varepsilon}{\partial \theta} = K_U \cdot \sin 2(\theta_e - \alpha) - K_f \cdot \sin 2\theta_e = 0 \qquad (5)$$

where θ_e is the angle of equilibrium between two anisotropies (shape and magnetostatic interaction) that was previously estimated as 35°. Solution of (5) gives the relation $K_f / K_u = f_V^2 / \Delta N = \sin 2(\theta_e - \alpha) / \sin 2\theta_e$ providing $\alpha = 15°$ that is in good agreement with the corresponding value estimated from TEM.

The balance equation $\partial \varepsilon / \partial \theta = 0$ for in-plane orientation of the external field H_{II} can be expressed as

$$\frac{\partial \varepsilon}{\partial \theta} = -K_u \cdot \sin 2(\theta + \alpha) + K_f \cdot \sin 2\theta + M_S \cdot H_{II} \cdot \sin \theta = 0 \qquad (4)$$

where α is the angle between NPs long axis and the film normal, now. Saturation of the system magnetization is achieved when $\theta = 0$ for any field exceeding H_a. From the stability condition $\partial^2 \varepsilon / \partial \theta^2 > 0$, angle θ equals zero in the case of

$$H_{II} > \frac{2(K_u \cos 2\alpha - K_f)}{M_S} = 4\pi \cdot M_S \cdot (\Delta N \cdot \cos 2\alpha - f_V^2) \qquad (5)$$

The latter gives estimated value for $H_a' = 4\pi \cdot M_S \cdot (\Delta N \cdot \cos 2\alpha - f_V^2)$ provided 1.4 kOe for $x = 73$ film that coincide well with the inflection point of $\mu(H_{II})$ curve (see Figure 5a). Estimation of the anisotropy field in the frame of this approach is in better agreement with the experimental data than H_a value obtained from (1), and takes additionally into account the deviation of NPs long axes from the film normal. Thus, basing on the obtained values for H_a' and α', one can estimate the influence of magnetostatic interaction between nanocolumns on the PMA of the studied system appearing in the decrease in anisotropy field induced by the shape of individual nanocolumns from $H_a = 4\pi \cdot \Delta N \cdot M_S \sim 8$ kOe to 1.5-2 kOe, as well as in tilting NPs easy magnetization axes from the films normal by ~20° under external field applied.

Finally, obtained experimentally and calculated by different procedures parameters characterizing magnetic anisotropy of α-FeCo(Zr) ferromagnetic nanocolumns in as-deposited FeCoZr-CaF$_2$ films are collected in Table 1. H_a is the anisotropy field, H_a' is the anisotropy field modified by NPs magnetic interaction, H_d is the demagnetizing field, α is the angle between easy axis of NPs magnetization and films normal, α+α' is the orientation of NPs easy axis (with respect to the films normal) modified by interaction under external field applied.

Table 1. Magnetic parameters characterizing PMA of α-FeCo(Zr) ferromagnetic nanoparticles in the as-deposited FeCoZr-CaF$_2$ films

		H_a, kOe	H_a', kOe	H_d, kOe	α, °	α+α', °
(FeCoZr)$_{58}$(CaF$_2$)$_{42}$						
from experimental curve/spectrum		—	~1.7	~8	15*-29**	—
(FeCoZr)$_{73}$(CaF$_2$)$_{27}$						
from experimental curve/spectrum		—	1.5	11.5	15*-24**	—
calculated	for prolate isolated spheroid	8	—	0	—	—
	interacting NPs (Néel approach)	—	3.5	9.7	—	—
	S&W model	—	0.5	—	—	35
	from energy balance equation	—	1.4	—	15	35

* - estimation from TEM images,
** - evaluation from line intensities in RT Mössbauer spectra.

PMA in Oxidized FeCoZr-CaF$_2$ Nanocomposite Films

As it is shown above, the most appropriate arrangement of ferromagnetic material in (FeCoZr)$_x$(CaF$_2$)$_{100-x}$ nanocomposites providing NPs columnar structure with the maximal PMA effect occurs in the film with the highest examined x value of 73 at.%. For decreasing NPs magnetostatic interaction that enlarges with x and masks PMA effect, partial oxidation of NPs is carried out by sputtering the films in the appropriate oxygen-containing ambient [4, 8]. Verification of NPs columnar structure as well as their partial oxidation is fulfilled by HRTEM and EELS. The results of the films structural analysis are presented in Figures 7 and 8. As it can be seen from cross-section TEM and HRTEM images, evidently elongated shape of metallic NPs, similar to that in non-oxidized films, occurs similarly in the film sputtered in oxygen-containing ambient. Additionally, cross-sectional diameter of columnar nanostructures does not change visibly as the result of oxygen adding in sputtering atmosphere. Namely, their diameter of 3-4 nm is very close to the size of non-oxidized NPs (see Figure 3b). However, flat contrast between NPs and matrix that separates them should be pointed out as compared with non-oxidized film. The reason is in partial oxidation of NPs previously detected in similar films [4, 8]. Indeed, formation of NPs oxide shells stated in [4, 5] leads to the decrease in average NPs density. In agreement with the observed elongated shape of NPs, calculation of demagnetizing factors of NPs in the oxidized film with $x = 73$ basing on its ferromagnetic-resonance spectra provides $N_{OP} = 0.04 \cdot 4\pi$ and $2 \cdot N_{IP} = 0.96 \cdot 4\pi$ values. They indicate that the shape of NPs is very close to the infinite cylinder, similarly to non-oxidized film.

Crystallinity of nanocolumns is confirmed by comparing bright-field and dark-field selected area TEM images (Figure 7, c and d). Bright elongated areas on dark-field image correspond to metallic nanocolumns on the respective bright-field image that indicates NPs crystalline structure. However, XRD pattern of the film studied (see Figure 7e) evidences only two significantly broad diffraction lines corresponding to α-FeCo(Zr) phase that indicates higher degree of disorder of NPs crystalline structure in oxidized sample as compared to non-oxidized film with the same composition (Figure 3b). Additionally, broad bump can be detected in the region of 2Θ~32° that corresponds possibly to iron oxide. Single broadened line does not allow oxide interpretation, but previous complex analysis of the film phase composition associate it with nearly amorphous hematite as the most probable candidate [4].

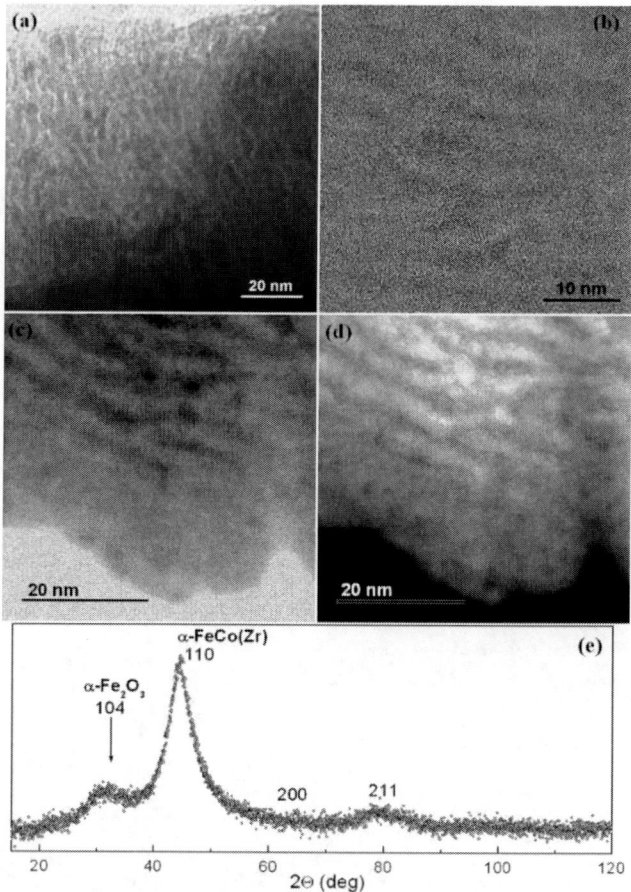

Figure 7. Cross-section TEM (a), HRTEM (b), selected area bright-field (c) and dark-field (d) images as well as XRD pattern (e) of $x = 73$ film sputtered in oxygen-containing atmosphere with $P_O = 4.3$ mPa.

Detailed elements mapping carried out by EELS for single NP in oxidized $x = 73$ film is presented in Figure 8. Dark-field TEM image detects elongated nanoparticle with crystalline central part. According to EELS spectra, Co dominates in central part of NP. Additionally, rather sharp contrast between cobalt and oxygen can also be detected at Co-O map that points out low cobalt oxidation degree previously detected by XANES and EXAFS spectroscopies [4]. In contrast, Fe is present both in central and in surface parts of NP, but Fe-Co map illustrates separation of Fe and Co and domination of Fe at NP surface. Additionally, there is no such a sharp contrast between Fe and O as in the case of Co that indicates at least partial Fe oxidation confirming previous XANES and EXAFS results [4]. Oxygen is evidently located at the surface of nanoparticle consistently with the "metallic core – oxide shell" model. Thus, combined TEM and EELS results prove formation of elongated columnar-like metallic NPs in the oxidized films, similar to that obtained in oxygen-free atmosphere, but covered with the oxide shells at their surface.

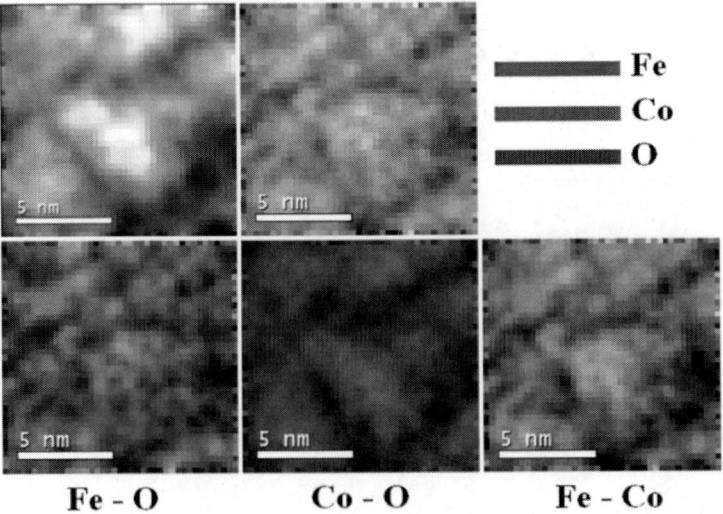

Figure 8. Dark-field TEM image accompanied by EELS spectra of selected area of $x = 73$ film sputtered in oxygen-containing atmosphere.

Magnetic anisotropy of the system of partially oxidized α-FeCo(Zr) nanocolumns is analyzed complementary by Mössbauer spectroscopy (Figure 9a) and magnetometry (Figure 9b). Mössbauer spectroscopy detects magnetic splitting of spectra of the studied film measured at RT and 78 K which points out NPs magnetic ordering. However, in contrast to non-oxidized film, two different types of Fe local ordering are detected corresponding to non-oxidized α-FeCo(Zr) phase (Fe^0, δ = 0.05 mm/s) and Fe^{3+} oxide (δ = 0.39 mm/s) proving partial NPs oxidation. The latter phase possesses broad sextet lines and wide B_{eff} distribution indicating high degree of local structure disordering because of its location at NPs surface. The ratio between non-oxidized Fe atoms and Fe^{3+} ions is 36: 64 according to Mössbauer spectra approximation. The relation of spectral lines intensities for sextet describing α-FeCo(Zr) phase is h_1:h_2:h_3 = 3:0.15:1 providing α = 16° at RT that is significantly lower than α = 24° for corresponding phase in non-oxidized film. This value is very close to the parameter α = 15° expected from TEM images (Table 1) and seems to originate from stochastic dispersion in the growth orientations of nanocolumns. Temperature decrease down to 78 K leads to 2° reduction of α value corresponding to decrease in thermal fluctuations, similarly to the case of non-oxidized film. Additionally, it should be mentioned that the ratio of spectral lines of sextet describing iron oxide is also characterized by low K value, less than 0.1, providing α = 11° that indicates almost collinear orientation of magnetic moments in both metallic cores and oxide shells of NPs in the direction close to the film normal.

The dependences of μ(H) of the oxidized x = 73 film measured in two orthogonal directions of external field, namely $H_∥$ and $H_⊥$, are presented in Figures 9b and 9c, respectively. The main peculiarities of μ(H) dependences characterizing oxidized film are in significant increase in H_a and decrease in H_d if compared with corresponding parameters of non-oxidized x = 73 film. As it can be seen from Figure 9b, H_a value (it should be defined correctly as H_a', because the parameter determined directly from μ($H_∥$) curve is modified by NPs interaction, as it is mentioned above) achieves 2 kOe at RT and 3.5 kOe at low temperatures (T = 2-100 K) that is more than two times higher than corresponding parameter of non-oxidized film. It coincides with mentioned before value predicted in the frame of Néel approach (1) for magnetically interacting nanocolumns. Previously obtained disagreement of theoretical prediction and experimental data for non-oxidized $x = 73$ film was explained by strong conglomeration of magnetic NPs and their higher interaction than Néel model assumes. In so doing, partial NPs oxidation seems to prevent such "extra-interaction" and promote achieving theoretically predicted H_a value characterizing anisotropy of "ideal" nanocolumns assembly with given geometrical (ΔN, f_V) and magnetic (M_S) parameters.

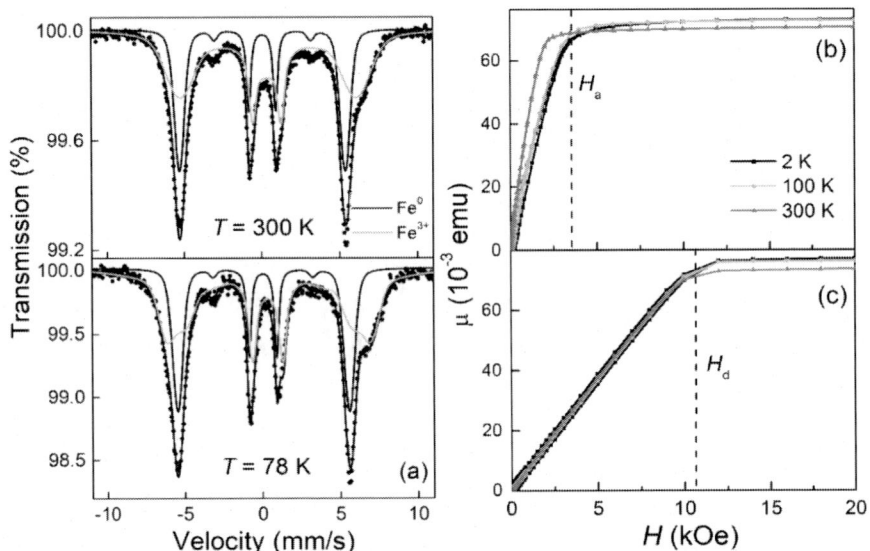

Figure 9. Mössbauer spectra (a) and $\mu(H)$ dependences (b, c) of partially oxidized $x = 73$ film measured in the film plane $H_{\parallel}$ (b) and in the film normal $H_\perp$ (c) directions at different temperatures. Dashed lines in (b) and (c) correspond to H_a and H_d values of the film at $T = 2$ K, respectively.

Similarly, H_d parameter sensitive to NPs interaction decreases down to 10.5 kOe for oxidized $x = 73$ film that is closer to the value of 9.7 kOe calculated according to $H_d = 4\pi \cdot M_S \cdot f_V$ relation than 11.5 kOe extracted from $\mu(H_\perp)$ curve of non-oxidized film. Additionally, it should be mentioned that evidently more sharp inflection points in $\mu(H_{\parallel})$ and $\mu(H_\perp)$ dependences characterizing saturation of NPs magnetization in hard and easy directions can be detected in the case of partially oxidized film comparing with the samples sputtered in Ar. Such a phenomenon is supposed to indicate decrease in the dispersion of NPs magnetic moments orientations as the result of NPs partial oxidation, in agreement with the results of Mössbauer spectroscopy.

Thus, basing on correlation of the structural and magnetic properties, one can conclude strong influence of the oxide shells formed around α-FeCo(Zr) columnar NPs on their PMA. Progress is associated rather with the role of oxide shells as additional nonmagnetic barriers preventing NPs interaction, because expected α-Fe_2O_3 oxide possesses weak ferromagnetic or antiferromagnetic properties in the range of applied temperatures. Direct indication of magnetic interaction breaking is in the decrease of H_d value. In turn, decrease in planar component of magnetic anisotropy promotes revealing the "intrinsic" (i.e., shape) anisotropy of magnetic nanocolumns that provides PMA of the films. Corresponding significant α decrease detected clearly by Mössbauer spectroscopy and substantial increase in H_a observed by magnetometry is supposed to be the consequence of such PMA "refinement."

One more key source of PMA enhancement, besides magnetostatic interaction preventing, is better arrangement of magnetic moments of α-FeCo(Zr) nanocolumns in the direction of film normal, i.e., along their long axes, in oxidized sample comparing with the non-oxidized film. Among the observations mentioned above and proving such extra-arrangement are transformation of μ(H) dependences shape after oxidation, which becomes more linear with sharp inflection points in the region of saturation (Figures 9, b and c) and drop of α from 24 to 16° detected by Mössbauer spectroscopy. The origin of such magnetic moments stabilization along the long axes of nanocolumns as the result of oxide shells formation is not clear evidenced. Inversely, non-oxidized central part of NPs possessing ferromagnetic properties becomes smaller that can induce superparamagnetic effects. One of possible interpretations of detected phenomenon is in positive influence of exchange coupling appearing between ferromagnetic core of NPs and their presumably antiferromagnetic oxide shell on PMA. Indeed, exchange coupling in "core-shell" nanostructures in partially oxidized FeCoZr-CaF$_2$ films with lower concentration of FeCoZr (x = 30-43 at.%) sputtered at P_O = 4.3 mPa was previously established basing on detected bias of field cooled (FC) $\mu(H)$ curves [4]. It should be mentioned that any shift of hysteresis loop is not detected for the studied x = 73 film in both $H_{\parallel}$ and $H_{\perp}$ in the whole range of measuring temperatures after field cooling in H_{cool} = 50 kOe (Figure 10). However, high values of coercive field H_C are detected at low temperatures (T = 2-100 K) achieving 226 and 254 Oe in $H_{\parallel}$ and $H_{\perp}$, respectively. For comparison, RT values of H_C amount 55 and 30 Oe. It should be mentioned that identical H_C values are obtained both after FC (Figure 10) and zero-field cooled (ZFC) (Figure 9, b and c) procedures of low temperature $\mu(H)$ loops measurement. Such low temperature increase in magnitude of H_C cannot be explained traditionally by superparamagnetic effects, because Mössbauer spectroscopy clearly detects magnetic ordering in the oxide even at RT (Figure 9a). Any comparable effect was not revealed in non-oxidized x = 73 film that demonstrates almost temperature-independent H_C increasing by 10-15 Oe only with T decrease (from 38 to 48 Oe in $H_{\parallel}$ and from 38 to 55 Oe in $H_{\perp}$). However, simple oxidation accompanied by formation of high-coercive oxides cannot be involved in H_C increase explanation, because (*i*) high H_C is detected only for low temperatures whereas magnetically ordered oxide should demonstrate almost temperatures-independent H_C (excepting phase transitions), and (*ii*) α-Fe$_2$O$_3$ oxide formation is expected which is antiferromagnetic bellow Morin transition temperature (T_M ~ 260 K for bulk hematite [11] and ~ 150 K for 27 nm nanoparticles [16]).

It should be emphasized that pronounced H_a increase up to 3.5 kOe for the oxidized film is detected at T = 2-100 K only, whereas RT value of H_a is 2 kOe. One more peculiarity of low temperature $\mu(H)$ dependences of the oxidized film is open loops in both field directions almost up to saturation. In common, above mentioned features of magnetization curves accompanied by Mössbauer data indicating preferential orientation of magnetic moments in oxide in the absence of external field point out that oxide shells organize itself during film sputtering consistently with the magnetic ordering in ferromagnetic core. In other words, surface oxidation of ferromagnetic NPs with strong magnetic anisotropy leads to formation of magnetically oriented oxides reproducing ferromagnetic cores preferential orientation. In turn, being coupled with antiferromagnetic oxide, ferromagnetic columnar core of NPs demonstrates more stable orientation of magnetic moments resulting in their higher anisotropy. In other words, antiferromagnetic oxide shells hold surface magnetic moments of ferromagnetic core along NPs easy axes preventing their fluctuations. Exclusively low-temperature increase in H_C correlates with the fact that Morin transition of hematite to antiferromagnetic state is bellow RT. Strong magnetic anisotropy of ferromagnetic nanocolumns probably exceeding the anisotropy of antiferromagnetic oxide shell explains zero µ(H) loop shift after field cooling [17]. The absence of additional H_C increase after FC as compared to ZFC µ(H) curve indicates that ferromagnetic nanocolumns produce enough magnetostatic field for oxide arrangement, that any additional FC is not necessary for oxide extra-ordering.

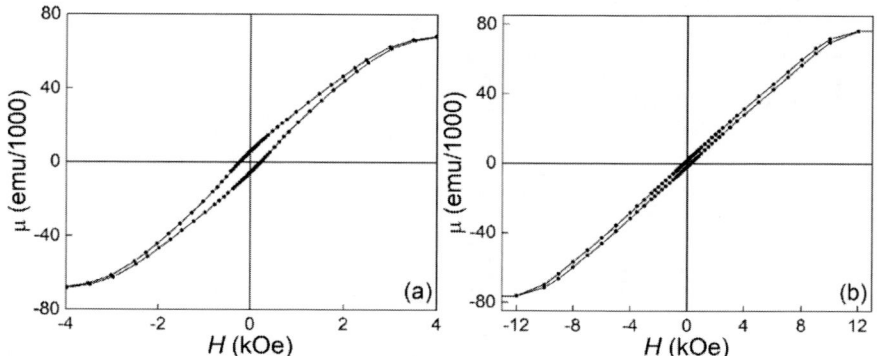

Figure 10. Parts of FC µ(H) dependences of partially oxidized $x = 73$ film measured at $T = 2$ K in the film plane $H_{\parallel}$ (a) and in the film normal $H_{\perp}$ (b) directions after cooling at $H_{cool} = 50$ kOe.

Irradiation of Non-oxidized FeCoZr-CaF$_2$ Nanocomposite Films

Heavy ions irradiation of the studied nanocomposite films aimed to increase their PMA is carried out along the films normal both for samples with non-oxidized and partially oxidized NPs. Wide range of Xe-ions fluences is applied for analysis correlation between treatment conditions and structural as well as magnetic properties induced in the films. TEM image of non-oxidized (FeCoZr)$_{73}$(CaF$_2$)$_{27}$ film irradiated with $D = 8 \cdot 10^{12}$ ion/cm^2 is presented in Figure 11a.

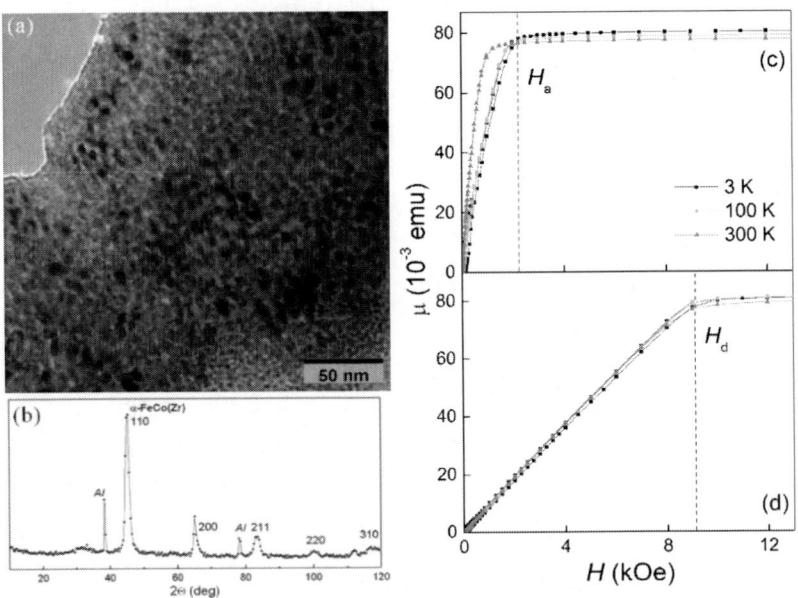

Figure 11. TEM image (a), XRD pattern (b) and μ(H) dependences (c, d) of non-oxidized $x = 73$ film irradiated with Xe ions of $D = 8 \cdot 10^{12}$ ion/cm^2. μ(H) dependences in (c) and (d) are measured in the film plane $H_{\parallel}$ and in the film normal $H_{\perp}$ directions, respectively. Dashed lines in (c) and (d) correspond to H_a and H_d values of the film at $T = 3$ K, respectively.

As can be seen from the Figure, the irradiated film contains elongated NPs similar to those in as-deposited (non-irradiated) film. Any ion track was not detected because of possible track overlapping. Precise estimation of NPs elongation is impossible from the Figure 11a, because the obtained image is not exactly cross-sectional to the film plane, unfortunately. However, it can be seen clearly that nanocolumns diameter and the distance between them was not changed markedly as the result of irradiation (compare Figures 3b and 11a).

Calculations basing on ferromagnetic-resonance spectroscopy confirm cylindrical shape of NPs with $N_{OP} = 0$ and $2 \cdot N_{IP} = 4\pi$ [3], similarly to NPs in non-irradiated film. However, obtained by this method value of H_d characterizing NPs separation demonstrates significant decrease of about 1.5 kOe (from 9.9 to 8.3 kOe) after irradiation with $D = 8 \cdot 10^{12}$ ion/cm^2 as compared with the initial film. Lower H_d indicates more effective NPs separation preventing their interaction as well as minimization of planar component of magnetic anisotropy. The possible reason of obtained decrease is in "cutting-back" of NPs by ion track or by structural demages in the regions of their overlapping. In other words, some NPs or their parts lose ferromagnetic properties in the regions, where heavy ions form destructions or structural transformations.

It should be emphasized, that structural analysis of irradiated film by XRD (Figure 11b) reveal single α-FeCo(Zr) phase in well-crystallized state, identical to that in the as-deposited film. Some reduction of peaks width from 1.4 to 1.0° as compared with non-irradiated film is even detected (the point is about FWHM of (110) peak of α-FeCo(Zr) phase). The latter indicates the absence of any crystallinity breaking in NPs expected as the result of irradiation. Similarly, Mössbauer spectroscopy does not reveal any changes in ferromagnetic local ordering in NPs after irradiation with $D = 8 \cdot 10^{12}$ ion/cm^2, as the spectrum (not shown) being similar to that of the initial film (Figure 4d). Calculations show, that NPs magnetic moments conserve the same average orientation ($\alpha \sim 24°$) as before irradiation. The only difference between obtained hyperfine parameters of the two films studied is in $<B_{eff}>$ decrease from 33.1 ± 0.2 T to 32.3 ± 0.2 T after irradiation. The latter can indicate however some local disturbance of magnetic order in NPs.

Surprisingly, significant changes in magnetic properties of irradiated film are detected basing on the analysis of its μ(H) dependences. As can be seen from Figures 11, (c) and (d), essential increase in H_a up to 2 kOe and decrease in H_d down to 9 kOe is characteristic for irradiated film as compared with the initial one (1.5 and 11.5 kOe, respectively). It should be noted, that H_a demonstrates increase only at low temperatures $T = 2\text{-}100$ K, whereas corresponding RT parameter of about 1 kOe is even lower than for the initial film. On the contrary, H_d is almost independent on temperature. It should be mentioned that H_d value obtained for irradiated film is the lowest among the previously studied films and is the closest to theoretically predicted parameter of 9.7 kOe according to $H_d = 4\pi \cdot M_S \cdot f_V$ relation. The drop of H_d parameter after irradiation correlates with data obtained by ferromagnetic-resonance spectroscopy [3] and indicates better separation between ferromagnetic columnar NPs in the treated film. Reduction of planar component of magnetic anisotropy can result in PMA enhancement confirmed by H_a increase. It is important to point out that H_a does not achieve however the value of corresponding parameter obtained previously for partially oxidized as-deposited film ($H_a = 3.5$ kOe). This fact clearly indicates that PMA enhancement in partially oxidized film is caused not only by separation of NPs by non-ferromagnetic oxide shells, but additionally by core-shell exchange interaction in NPs.

For more pronounced PMA improving, higher Xe fluence of $1 \cdot 10^{13}$ ion/cm² is applied. Cross-sectional TEM image of irradiated $x = 73$ film is presented in Figure 12a. As can be seen from the Figure, NPs look now as the elongated chains of ellipsoidal or spherical nanostructures oriented in different directions. Thus, columnar ordering is destroyed by irradiation according to TEM analysis. Corresponding widening of diffraction lines is obtained by XRD evidencing in increase of FWHM of (110) peak up to 6° (Figure 12b). It should be mentioned that just Lorentzian component responsible for crystallites size provide FWHM increase according approximation by Rietveld procedure. The latter point out NPs size decrease that correlates with TEM results presenting nanocolumns division onto separate NPs. Contrary, better CaF_2 matrix crystallization can be stated basing on XRD results.

Concerning magnetization properties, H_a and H_d parameters remain almost unchangeable comparing to the corresponding values characterizing non-irradiated film.

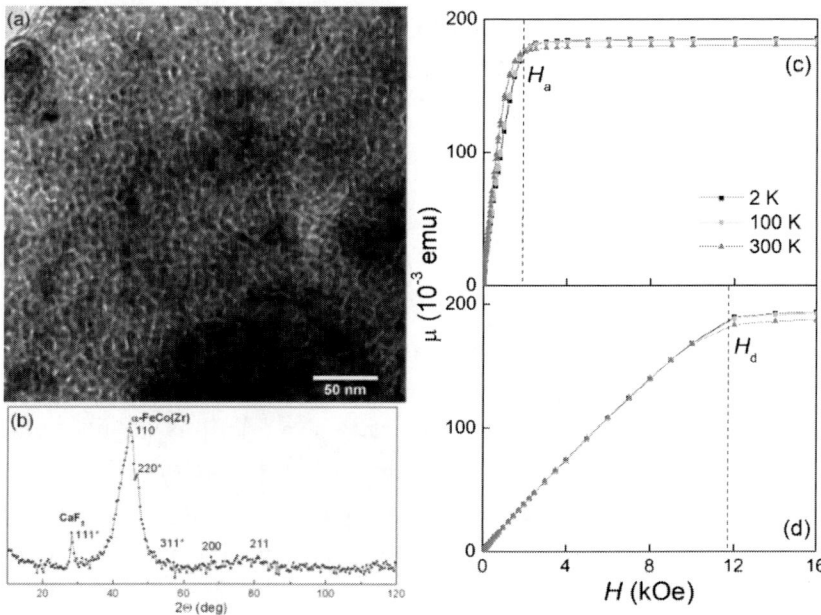

Figure 12. TEM image (a), XRD pattern (b) and μ(H) dependences of non-oxidized x = 73 film irradiated with Xe ions of $D = 1 \cdot 10^{13}$ ion/cm^2; μ(H) curves in (c) and (d) are measured in the film plane $H_\parallel$ and in the film normal $H_\perp$ directions, respectively.

They achieve 1.8 and 11.5 kOe, respectively (Figures 12, c and d). Small increase in H_a (comparing with 1.5 kOe for the initial film) is obtained only at low temperatures (2-100 K), whereas RT value $H_a = 1.3$ kOe is similar to as-deposited film. The results of magnetometry are in a good agreement with ferromagnetic-resonance spectroscopy data [3] indicating even some increase in H_d comparing with initial film from 9.9 to 11.5 kOe. Additionally, slight deviation of the shape of magnetic NPs from infinite cylinder to elongated ellipsoid is detected in terms of granular demagnetizing factors $N_{OP} = 0.04 \cdot 4\pi$ and $2 \cdot N_{IP} = 0.96 \cdot 4\pi$. Inversely, Mössbauer spectroscopy shows however decrease in α down to 21° (see Figure 13a) if compared with initial film (α ~ 24° in Figure 4b). Detectable increase in spectral line width after irradiation accompanied by reduced $<B_{eff}> = 32.4$ T is in agreement with XRD data (Figure 12b) and indicates distribution in NPs sizes. Summarizing magnetic properties of the film irradiated with $D = 1 \cdot 10^{13}$ ion/cm^2 obtained by different methods, there is no evident changes in anisotropy of the film after irradiation in spite of structural changes revealed by TEM and XRD. Thus, irradiation with Xe fluence of $D = 1\cdot 10^{13}$ ion/cm^2 influences dramatically on the film nanostructure inducing nanocolumns splitting, but however conserves its magnetic anisotropy oriented mainly perpendicular to the film plane, as in the initial film. Unfortunately, PMA of irradiated film is masked by increased disordered and corresponding thermal fluctuations probably induced by superparamagnetic effects. Thus, PMA is evident mostly at low temperatures (by magnetometry) or by methods with low characteristic time (i.e., Mössbauer spectroscopy).

Comparative analysis of PMA effect in $x = 73$ films obtained under different sputtering and treatment conditions can be simply illustrated by they $\mu(T)$ dependences measured in field cooled (FC) and zero-field cooled (ZFC) regimes (Figure 13b). It was previously shown [1], that NPs magnetization measured in their hard axis direction (i.e., in film plane) in low magnetic field reduces with temperature decrease. The origin of such reduction is in thermal fluctuation decrease. Indeed, low magnetic field cannot overcome the influence of shape anisotropy, which enforces NPs to orient their magnetic moments in the direction orthogonal to the external field. As can be seen from Figure 13b, comparison of $\mu(T)$ curves presented in arbitrary units for different films reflect described above tendency – the enhancement of PMA as the result of NPs partial oxidation as well as films irradiation. The worst results of PMA improvement are characteristic for irradiation with higher Xe fluence of $1\cdot 10^{13}$ ion/cm^2. In this case, anisotropy changes slightly, but small positive influence can be detected. The best results can be obtained using partial oxidation of NPs as well as by the film irradiation with lower fluence of $8\cdot 10^{12}$ ion/cm^2. Surprisingly, that the latter method of PMA improvement shows better result according to $\mu(T)$ curves analysis, whereas higher H_a is characteristic, in opposite, for the oxidized film. However, the unique result of maximal H_d decrease is obtained exactly for the film irradiated with $D = 8\cdot 10^{12}$ ion/cm^2. It should be mentioned, that $\mu(T)$ curves measured along NPs long axes (i.e., along the films normal, not shown) remain almost unchangeable with T for all the samples studied because of strong NPs shape anisotropy suppressing thermal fluctuations effectively.

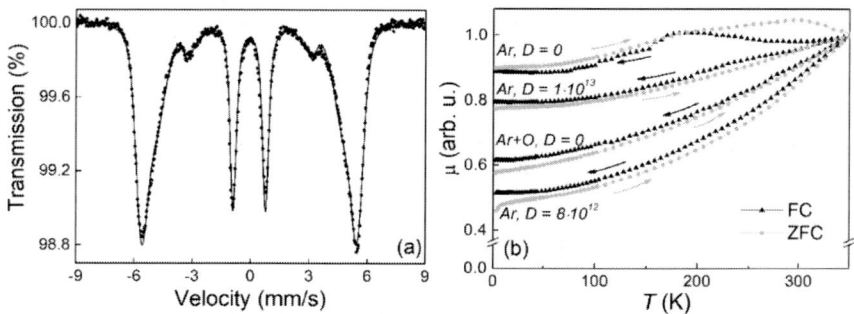

Figure 13. Mössbauer spectrum of non-oxidized $x = 73$ film irradiated with Xe ions of $D = 1 \cdot 10^{13}$ ion/cm^2 (a) and FC-ZFC $\mu(T)$ dependences measured in $H = 50$ Oe in the films plane (b) for $x = 73$ films obtained under different sputtering and treatment conditions.

Approximation of $\mu(H_\perp)$ dependences that was measured at $T = 2$ K for $x = 73$ films irradiated with both Xe fluences in the frame of Stoner-Wohlfarth model after the curves correction in the assumption of self-consistent fields (2)-(3) described above is presented in Figure 14. Considerable differences in magnetization reversal process are demonstrated by NPs in these two films when magnetized in the direction close to their long axes. Estimated parameters of H_a and α obtained from the fit amount to 1.9 kOe & 10° for the film irradiated with $D = 8 \cdot 10^{12}$ ion/cm^2 and 0.7 kOe & 30° for the film irradiated with $D = 1 \cdot 10^{13}$ ion/cm^2. In spite of the fact, that H_a parameter is underestimated for the latter film, it gives good agreement with H_a evaluated from $\mu(H_{\text{II}})$ curve for $D = 8 \cdot 10^{12}$ ion/cm^2 regime. Additionally, obtained from the fit parameters clearly illustrate tendency of H_a decrease with the growth of Xe fluence. The shape of magnetization curve (refined from the demagnetization effects) of the film irradiated with $D = 8 \cdot 10^{12}$ ion/cm^2 (Figure 14a) almost ideally corresponds to magnetization reversal along the easy magnetization axis, whereas $\mu(H_\perp)$ dependence of the film irradiated with $D = 1 \cdot 10^{13}$ ion/cm^2 indicates significant tilting the external field orientation with respect to the easy axis direction. The latter correlates with TEM image of the film (Figure 12a) and is the consequence of NPs shape transformation. Dispersion in orientations of NPs as well as the decrease in their elongation results in more pronounced magnetic moments dispersion and subsequent increase in the average angle of their deviation from the film normal.

Irradiation of Partially Oxidized FeCoZr-CaF$_2$ Films

Basing on the above obtained results presenting positive influence of partial oxidation and irradiation with low fluences on PMA of (FeCoZr)$_{73}$(CaF$_2$)$_{27}$ films, combination of these two methods is believed to be prospective procedure appropriate for simultaneous H_a increase and H_d decrease that was achieved previously only separately from each other. In other words, the most effective H_a increase was achieved by NPs partial oxidation accompanied by "ferromagnetic core – antiferromagnetic shell" structures formation, whereas the most pronounced decrease in H_d was carried out by the film irradiation promoting NPs separation probably as the result of destructions induced by ion tracks.

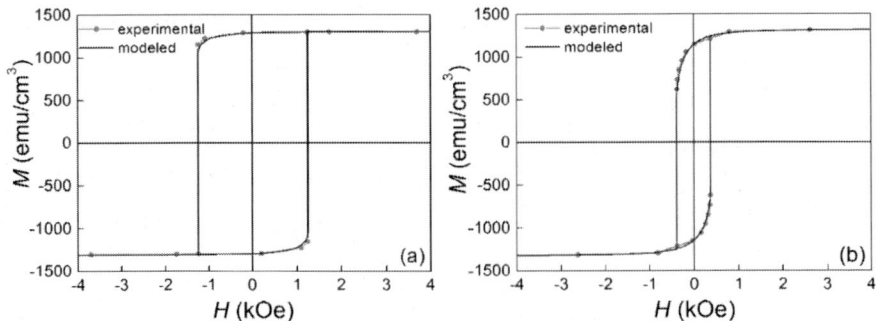

Figure 14. Parts of $\mu(H_\perp)$ dependences measured at 2 K for non-oxidized $x = 73$ films irradiated by Xe ions with $D = 8 \cdot 10^{12}$ ion/cm^2 (a) and $1 \cdot 10^{13}$ ion/cm^2 (b), accompanied by their approximation in the frame of Stoner-Wohlfarth model after refinement from demagnetizing effect.

The results of structural analysis of $x = 73$ films sputtered in oxygen containing atmosphere ($P_O = 4.3$ mPa) and subsequently irradiated by Xe ions with the fluences variable in the wide range of $D = 5\text{-}25 \cdot 10^{12}$ ion/cm² are presented in Figure 15a.

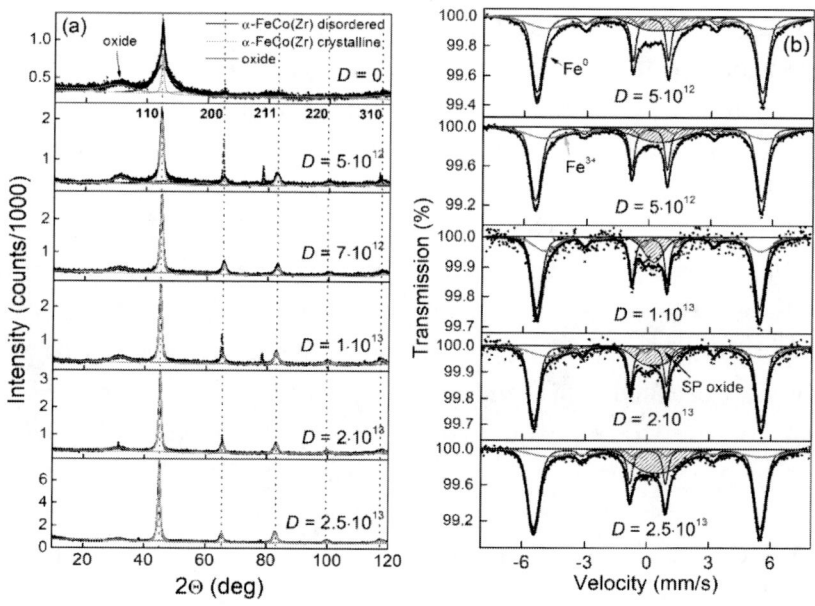

Figure 15. XRD patterns (a) and Mössbauer spectra (a) of partially oxidized $x = 73$ films irradiated by Xe ions with varied fluences D (the values are given in ion/cm²).

As can be seen from the Figure, significant narrowing of diffraction lines accompanied by the appearance of additional peaks characterizing α-FeCo(Zr) structure is detected in the whole range of applied fluences. The width of (110) peak decreases gradually with D increase from 1.1° for $D = 5 \cdot 10^{12}$ ion/cm² through 1.0° for $D = 7\text{-}10 \cdot 10^{12}$ ion/cm² and 0.9° for $D = 20 \cdot 10^{12}$ ion/cm² to 0.8° for $D = 25 \cdot 10^{12}$ ion/cm². This observation clearly indicates that irradiation of the oxidized films improves crystalline structure of NPs cores which become similar to that in non-oxidized film (see Figure 3b). Moreover, some sharpening the bump characterizing oxide shell is also detected. This tendency is replaced then by decrease in the intensity of line corresponding to oxide with D growth. The detected phenomenon of monotonic crystalline structure improvement differs oxidized films from the corresponding non-oxidized films which demonstrate the improvement of α-FeCo(Zr) crystallinity only for $D = 8 \cdot 10^{12}$ ion/cm^2, whereas higher fluence of $D = 1 \cdot 10^{13}$ ion/cm^2 introduce disordering in NPs crystalline structure.

As can be seen from the Figure, significant narrowing of diffraction lines accompanied by the appearance of additional peaks characterizing α-FeCo(Zr) structure is detected in the whole range of applied fluences. The width of (110) peak decreases gradually with D increase from 1.1° for $D = 5 \cdot 10^{12}$ ion/cm^2 through 1.0° for $D = 7\text{-}10 \cdot 10^{12}$ ion/cm^2 and 0.9° for $D = 20 \cdot 10^{12}$ ion/cm^2 to 0.8° for $D = 25 \cdot 10^{12}$ ion/cm^2. This observation clearly indicates that irradiation of the oxidized films improves crystalline structure of NPs cores which become similar to that in non-oxidized film (see Figure 3b). Moreover, some sharpening the bump characterizing oxide shell is also detected. This tendency is replaced then by decrease in the intensity of line corresponding to oxide with D growth. The detected phenomenon of monotonic crystalline structure improvement differs oxidized films from the corresponding non-oxidized films which demonstrate the improvement of α-FeCo(Zr) crystallinity only for $D = 8 \cdot 10^{12}$ ion/cm^2, whereas higher fluence of $D = 1 \cdot 10^{13}$ ion/cm^2 introduce disordering in NPs crystalline structure.

Investigation of the local structure and magnetic moments ordering in partially oxidized and subsequently irradiated films by Mössbauer spectroscopy (Figure 15b) allows detecting slight decrease in the intensity of the second and the fifth spectral lines after irradiation. Such changes are particularly visible for the films irradiated with $D = 7 \cdot 10^{12}$ and $2 \cdot 10^{13}$ ion/cm^2. This indicates that irradiation provides some further decrease in dispersion angle α induced previously by NPs partial oxidation. Calculation of α values basing on Mössbauer spectra fitting was carried out here by averaging corresponding parameter for all sextets detected in the film (describing α-FeCo(Zr) and oxide subsystems). The data obtained for the films irradiated with different fluences being almost the same (α = 21-24°). The lowest value was obtained for $D = 2 \cdot 10^{13}$ ion/cm^2 amounting to 16°. Additionally, α-FeCo(Zr) phase demonstrates spectral line narrowing after irradiation that is in agreement with XRD data. Simultaneously, contribution of subspectra characterizing non-oxidized

NPs cores increases from 36% for as-deposited film to 64% maximally in the film treated with $D = 1 \cdot 10^{13}$ ion/cm^2. Corresponding reduction of the sextet characterizing α-Fe$_2$O$_3$ phase (from 45 to 20%) is accompanied by the appearing superparamagnetic singlet (or doublet) enlarging gradually with D up to 23% for $D = 2.5 \cdot 10^{13}$ ion/cm^2. The latter indicates the reduction in the size of NPs oxide shells. Consistently, both decrease in NPs oxidation degree and improvement of α-FeCo(Zr) crystalline structure after irradiation in the wide range of applied fluences allows to conclude that irradiation release NPs from defects and dissolved oxygen, whereas oxide shells demonstrate a protective function preventing NPs destruction and division into small parts, as in the case of non-oxidized films (Figure 12a).

Magnetization curves of partially oxidized $x = 73$ films irradiated with $D = 7 \cdot 10^{12}$ ion/cm^2 and $2 \cdot 10^{13}$ ion/cm^2, which demonstrate better results of NPs magnetic moments ordering according to Mössbauer spectroscopy are presented in Figure 16.

As it can be seen from the Figure 16, a and b, H_a parameter estimated from $\mu(H_{\parallel})$ curves drops significantly after irradiation in both regimes down to 2.6 and 2.4 kOe (at $T = 2{-}100$ K), respectively, as compared with partially oxidized non-irradiated film ($H_a = 3.5$ kOe). It should be mentioned, that RT value remains almost unchanged (1.6-1.8 kOe if compared with 2 kOe for the initial film). However, H_a values of the partially oxidized films irradiated in both regimes exceed corresponding parameter of non-oxidized films, both initial and irradiated. Thus, detected decrease in H_a is associated with decrease in the size of NPs oxides shells revealed by Mössbauer spectroscopy which are responsible for high H_a values due to exchange coupling between highly anisotropic ferromagnetic cores and antiferromagnetic shells.

Analysis of $\mu(H_{\perp})$ curves of partially oxidized films after irradiation indicates clear difference in H_d value for two regimes of irradiation. Irradiation with $D = 7 \cdot 10^{12}$ ion/cm^2 does not induce any detectable variations of H_d if compared with as-deposited film ($H_d \sim 11$ kOe), whereas higher fluence of $D = 2 \cdot 10^{13}$ ion/cm^2 provides significant decrease in H_d that achieves 8 kOe after Xe treatment. The obtained value of H_d is even substantially lower than for irradiated non-oxidized films (9.2 kOe). Such a rapid decrease in H_d of the oxidized film after irradiation may be explained, besides non-magnetic ion tracks formation, by some transformations in oxide under irradiation, e.g., destructions and amorphization that eliminate antiferromagnetism. The latter can decrease H_a but enlarge nonmagnetic layer between ferromagnetic NPs, because antiferromagnetic shells can participate in indirect interaction between NPs cores. Significant H_d decrease after irradiation is additionally confirmed for the oxidized film by ferromagnetic-resonance spectroscopy, the decrease being gradual with D. First, H_d value decreases from 11 kOe for as-deposited film to 8.2 kOe after irradiation with $D = 7 \cdot 10^{12}$ ion/cm^2, and then it drops to 6.4 kOe after ion treatment with $D = 2 \cdot 10^{13}$ ion/cm^2. The tendency is in a good agreement with the magnetometry results.

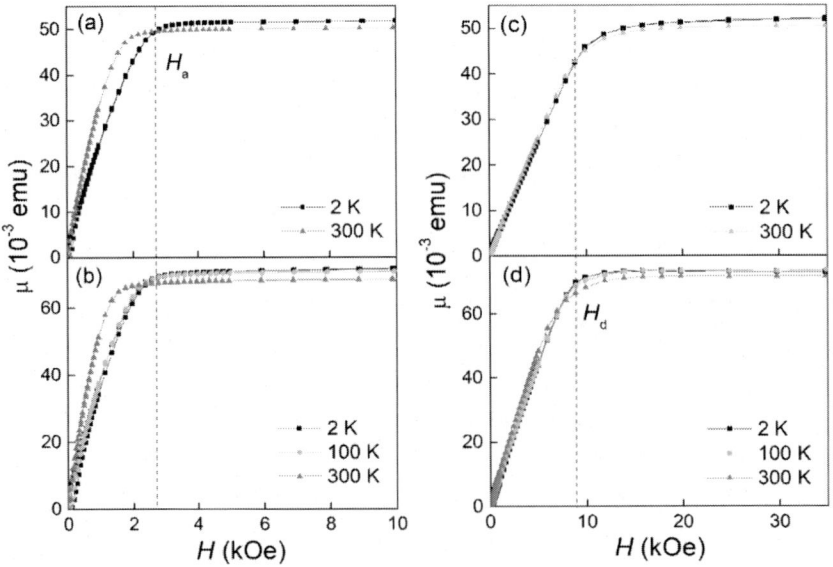

Figure 16. $\mu(H)$ dependences of partially oxidized $x = 73$ films irradiated by Xe ions with $D = 7 \cdot 10^{12}$ ion/cm^2 (a, c) and $2 \cdot 10^{13}$ ion/cm^2 (b, d), which are measured in the films plane (a, b) and in the films normal direction (c, d) at different temperatures. Dashed lines in (a-b) and (c-d) correspond to H_a and H_d values of the film irradiated with $D = 2 \cdot 10^{13}$ ion/cm^2, which are obtained at $T = 2$ K, respectively.

Table 2. Magnetic parameters characterizing perpendicular anisotropy of $(FeCoZr)_{73}(CaF_2)_{27}$ films sputtered in oxygen-free (Ar) and oxygen-containing (Ar+O) atmosphere, both as-deposited and subsequently irradiated by Xe ions with different fluences D

	H_a', kOe		H_d, kOe		α, °
	2 K	300 K	2 K	300 K	300 K
treatment condition	Ar sputtering ambient				
D = 0 (as-deposited)	1.5	1.3	11.5	11.3	24
D = 8·1012 ion/cm2	2.0	0.9	9.1	9.2	24
D = 1·1013 ion/cm2	1.8	1.3	11.6	11.0	21
	Ar+O sputtering ambient (PO = 4.3 mPa)				
D = 0 (as-deposited)	3.5	2.0	11.0	10.4	16
D = 7·1012 ion/cm2	2.6	1.8	10.9	10.4	22
D = 2·1013 ion/cm2	2.4	1.6	8.0	8.4	16

H_a' is the anisotropy field modified by NPs magnetic interaction, H_d is the demagnetizing field, α is the angle between easy magnetization axis of NPs and the film normal. Finally, the most important parameters characterizing magnetic anisotropy of $x = 73$ nanocomposite films improved by their partial oxidation and Xe^+ ions treatment are collected in the Table 2 for visualization of the observed phenomena.

CONCLUSION

In the presented work, one has carried out the complimentary study of structural and magnetic properties of $FeCoZr-CaF_2$ nanocomposite films demonstrating perpendicular magnetic anisotropy (PMA) originating from the shape anisotropy of ferromagnetic nanoparticles (NPs) elongated in the direction close to the films normal. Particular attention was concentrated on NPs partial oxidation and heavy ions irradiation as the methods of the enhancement the films PMA.

Basing on the results of Mössbauer spectroscopy and EELS, specific structure of NPs including "ferromagnetic α-FeCo(Zr) core – antiferromagnetic α-Fe_2O_3 shell" is proved for the films sputtered in oxygen-containing ambient with P_O = 4.3 mPa. Partial oxidation NPs turned out to be selective with respect to Fe, Co and Zr atoms implying preferential Fe and Zr oxidation with their dislocation mainly to the shell regions of NPs and almost non-oxidized state of Co in the core parts of NPs. Importantly, such partial oxidation of NPs in $(FeCoZr)_{73}(CaF_2)_{27}$ film leads to increase its PMA accompanied by the corresponding significant increase in the anisotropy field H_a up to 3.5 kOe if compared with non-oxidized film (H_a = 1.5 kOe). Obtained PMA enhancement is believed to originate from exchange coupling between metallic core and oxide shell in NPs promoting NPs magnetic moments stabilization along their long axes. The latter is proved by Mössbauer spectroscopy detecting the significant decrease in the dispersion of NPs magnetic moments with respect to the film normal from 24° to 16° as the result of NPs partial oxidation.

Irradiation of non-oxidized $(FeCoZr)_{73}(CaF_2)_{27}$ film with heavy Xe^+ ions promotes separation of ferromagnetic columnar NPs from each other by ion tracks, which are supposed to be non-magnetic. This leads to preventing NPs magnetostatic interaction and, as the result, decrease in demagnetizing field H_d of the film down to 9 kOe from the value of 11.5 kOe characterizing as-deposited film. Such a decrease in planar component of the film anisotropy assists to PMA contribution growth. Unfortunately, applying higher Xe^+ fluences (D = $1 \cdot 10^{13}$ ion/cm^2) starts to destroy columnar shape of NPs implying their division into separate ellipsoidal or spherical NPs, according to TEM data supported by XRD-based estimations.

Irradiation of preliminary oxidized $(FeCoZr)_{73}(CaF_2)_{27}$ film reveals that oxides shells protect NPs from destruction even in the case of high fluences applied (up to D = $2.5 \cdot 10^{13}$ ion/cm^2). Moreover, ion treatment of partially oxidized films releases NPs from defects and dissolved oxygen. The former improves NPs crystallinity evidenced from XRD data. The latter is associated with the decrease in the thickness of oxide shells in NPs confirmed by Mössbauer spectroscopy. This, in turns, leads to undesirable decrease in H_a down to 2.4 kOe, as compared with 3.5 kOe for non-irradiated film, due to the reduction of exchange coupling between metallic cores and oxide shells of NPs positively influencing on PMA. However, substantial decrease in H_d is achieved as the result of irradiation of the oxidized film reaching minimal obtained value of 8 kOe that indicates the weakest magnetic interaction between NPs due to the combined effect of the oxide shells and ion tracks.

REFERENCES

[1] Kasiuk, J. V., Fedotova, J. A., Przewoznik, J., Zukrowski, J., Sikora, M., Kapusta, Cz., Grce, A., and Milosavljevic, M. (2014). Growth-induced non-planar magnetic anisotropy in FeCoZr-CaF$_2$ nanogranular films: Structural and magnetic characterization. *J. Appl. Phys.*, 116, 044301.

[2] Timopheev, A. A., Bdikin, I., Lozenko, A. F., Stognei, O. V., Sitnikov, A. V., Los, A. V., and Sobolev, N. A. (2012). Superferromagnetism and coercivity in Co-Al$_2$O$_3$ granular films with perpendicular anisotropy. *J. Appl. Phys.* 111, 123915.

[3] Kasiuk, J., Baev, V., Fedotova, J., Skuratov, V., Bondariev, V., Żukowski, P., and Koltunowicz, T. N. (2015). Characterization of ion-induced changes in magnetic anisotropy of FeCoZr-CaF$_2$ nanocomposite films by resonance methods. *Electrical Review.*, 11, 280-283.

[4] Kasiuk, J., Fedotova, J., Przewoznik, J., Sikora, M., Kapusta, Cz., Żukrowski, J., Grce, A., and Milosavljević, M. (2016). Oxidation controlled phase composition of FeCo(Zr) nanoparticles in CaF$_2$ matrix. *Materials Characterization*, 113, 71-81.

[5] Fedotova, J. A., Przewoznik, J., Kapusta, Cz., Milosavljevic, M., Kasiuk, J. V., Zukrowski, J., Sikora, M., Maximenko, A. A., Szepietowska, D., and Homewood, K. P. (2011). Magnetoresistance in FeCoZr–Al$_2$O$_3$ nanocomposite films containing "metal core – oxide shell" nanogranules. *J. Phys. D: Appl. Phys.*, 44, 495001.

[6] Kasiuk, J., Fedotova, J., Przewoznik, J., Kapusta, Cz., Skuratov, V., Milosavljevic, M., Bondariev, V., and Koltunowicz, T. (2015). Ion-induced modification of structure and magnetic anisotropy in

granular FeCoZr-CaF$_2$ nanocomposite films. *Acta Physica Polonica*, 128, 828-831.

[7] Kasiuk, J., Fedotova, J., Przewoznik, J., Kapusta, Cz., Skuratov, V., Svito, I., Bondariev, V., and Koltunowicz, T. (2017). Ion irradiation of oxidized FeCoZr-CaF$_2$ nanocomposite films for perpendicular magnetic anisotropy enhancement. *Acta Physica Polonica A*, 132, 206-209.

[8] Kasiuk, J. V., Fedotova, J. A., Koltunowicz, T. N., Zukowski, P., Saad, A. M., Przewoznik, J., Kapusta, Cz., Zukrowski, J., and Svito, I. A. (2014). Correlation between Local Fe States and Magnetoresistivity in Granular Films Containing FeCoZr Nanoparticles embedded into Oxygen-free Dielectric Matrix. *J. Alloys and Compounds*, 586, S432–S435.

[9] Rodriguez-Carvajal, J. (1993). Recent advances in magnetic structure determination by neutron powder diffraction. *Physica B*, 192, 55–69.

[10] Rancourt, D. G., and Ping, J. Y. (1991). Voigt-based methods for arbitrary-shape static hyperfine parameter distribution in Mössbauer spectroscopy. *Nucl. Instr. Meth. B*, 58, 85–97.

[11] Bødker, F., Hansen, M. F., Koch, C. B., Lefmann, K., and Mørup, S. (2000). Magnetic properties of hematite nanoparticles. *Phys. Rev. B*, 61, 6826-6838.

[12] Wohlfarth. E. P. (1955). The Effect of Particle Interaction on the Coercive Force of Ferromagnetic Micropowders. *Proc. R. Soc. Lond. A*, 232, 208-227.

[13] Ryabchenko, S. M., Timopheev, A. A., Kalita, V. M., Stognei O. V., and Sitnikov, A. V. (2011). Features of ferromagnetic resonance in nanogranular films with perpendicular anisotropy of particles. *J. Appl. Phys.*, 109, 043903.

[14] Stoner, E. C., and Wohlfarth, E. P. (1948). A Mechanism of Magnetic Hysteresis in Heterogeneous Alloys. *Phil. Trans. R. Soc. A*, 240, 599-642.

[15] Cullity, B. D., and Graham, C. D. (2009). *Introduction to magnetic materials*, 2-nd edition. Hoboken, New Jersey: Wiley-IEEE Press.

[16] Bodker, F., and Morup, S. (2000). Size dependence of the properties of hematite nanoparticles. *Europhys. Letters*, 52, 217-223.
[17] Nogues, J., Sort, J., Langlais, V., Skumryev, V., Surinach, S., Munoz, J. S., and Baró, M. D. (2005). Exchange bias in nanostructures. *Phys. Rep.*, 422, 65–117.

In: A Closer Look at Magnetic Anisotropy
Editor: Georges Fremont

ISBN: 978-1-53617-566-0
© 2020 Nova Science Publishers, Inc.

Chapter 3

COLLOIDAL MAGNETIC FLUIDS: A SPECIAL CASE OF MAGNETIC ANISOTROPY

Daniel Mayer[*] *and Petr Polcar*[†]
Department of Theory of Electrical Engineering,
University of West Bohemia in Pilsen, Pilsen, Czech Republic

ABSTRACT

Magnetic fluids (thus nanacomposite magnetic material) represent relatively innovative and perspective material for many industrial applications. Recent rapid development of nanotechnologies allowed production of magnetic fluids with wide range of required physical and chemical properties.

One of characteristic attributes of magnetic fluids is the significant change of their physical properties in dependence on the application of the external magnetic field. In technical practice, the change of viscosity of these fluids is most commonly used, but other physical properties (e.g., magnetic permeability and dielectric permittivity) change as well.

---
[*] Corresponding Author's E-mail: mayer@kte.zcu.cz.
[†] E-mail:polcarp@kte.zcu.cz.

Magnetic fluids are significantly non linear and strongly anisotropic medium. This must be respected when designing different applications by the use of special mathematical apparatus. From the electromagnetic field point of view, the characteristic parameter of the magnetic fluid is the tensor of the magnetic permeability/dielectric permittivity. Presented work describes and innovate method of determination elements of this tensor by measurement carried out on the sample of magnetic fluid.

Further, the mathematic-physical properties of magnetic fluid as the medium with and orthogonal anisotropy during the time-varying magnetization are examined. Dissipative phenomena manifesting themselves as energy losses, which significantly impact the design of various electrical appliances, are then discussed.

Moreover, new levitation phenomena in magnetic fluids are discussed. The possibility to levitate bodies in magnetic liquid can lead to significant practical applications.

Several examples of perspective technical applications of magnetic liquids are presented as well. Three examples are discussed more in detail: controlled torsion damper for application in transportation, next ferrofluid controlled capacitor for application in sensor technology and peristaltic pump that takes advantage of magneto-elastic properties and shows some significant advantages compared to classical pumps design.

Keywords: magnetic anisotropy, ferromagnetic liquid, ferrofluid, magnetorheology

INTRODUCTION

In the 19th century some physicists (M. Faraday, T. J. Seebeck and others) verified their presumptions of the qualities of the magnetic field using liquids where fine metal dust was dissolved. A disadvantage of this environment was its instability - under the influence of gravity the dust tended to settle down. After more than 100 years' physicists reopened the issue and found out that these liquids can be stable if the dust particles are very fine. With extremely fine ferromagnetic particles their sedimentation occurs after a longer time. In these liquids the magnetic viscosity phenomenon was discovered it means that when they are exposed to the magnetic field their viscosity increases. These liquids were labeled as magnetic rheological. Soon production technology was developed, their physical - chemical qualities were examined and their applications were searched for in technical, medical and biochemical practice. The 1st patents for using ferrofluids were gained by Jacob Rabinow [1] in 1940.

In the 50s and 60s of the 20th century research in ferrofluids was embargoed. Since 1970s the knowledge was disclosed and many papers in journals and books have been published. Many interesting and useful applications have been realized and patented. Complex mathematic-physical theories have been described that provide information about structure and behavior of ferrofluids in stationary and dynamic state. The solution of magnetic fields (electro-magnetic, thermal power etc.) is difficult in systems containing ferrofluids as they are in a very non-linear and anisotropic environment. Although many materials were published on ferrofluids, the development and mainly usage of these prospective materials is not fully exploited yet. Contemporary detailed knowledge about the ferrofluids is dealt with in works [1 – 11].

PHYSICAL-CHEMICAL PRINCIPLE OF FERROFLUIDS

Ferrodluid Structure

Ferrofluids are permanently stable colloid suspensions of ferromagnetic particles in carrier liquid [12]. Suspension stability here means the quality that the suspension remains permanently homogenous, it is the ferromagnetic particles do not separate from the carrier liquid and do not settle at the bottom of the container (do not sediment), neither creates mutual aggregations. To reach stability ferrofluids must contain ferromagnetic particles of the size 5 to 15 nm (1 nm = 10^{-9}m), so called *nanoparticles*. Nanoparticles are usually formed by one Weiss domain. As a result of spontaneous magnetization it has a magnetic moment and represents an elementary magnetic dipole. Elementary dipoles influence each other. To prevent their aggregation, they are covered with stabilizer, i.e., a polymerous (macromolecular) coating, so called *detergent* formed by the chains of polar molecules (e.g., fatty acid), long 1 to 2 nm. Every chain is at one end bound with a nanoparticle and at the other end loosely attracted by the molecules of the carrier medium, Figure 1. Detergent is thus a surface active material that prevents direct contact between nanoparticles, causes repulsive forces between them and so prevents their aggregation.

The most common materials for ferromagnetic nanoparticles are magnetite (Fe_3O_4), maghemite (Fe_2O_2), cobalt (Co), iron (Fe) or iron nitride Fe_xN). The carrier liquid can be water, various oils, usually syntactic on hydrocarbon base, glycol and their compounds. Typical magnetic liquid contains (in volume): 5% ferromagnets, 10% detergent and 85% carrier liquid. Its relative permeability $\mu_r \approx 5$ drops with temperature and at Curie temperature gets the value $\mu_r = 1$, saturation magnetization is about 1,3 T and working temperature is from −125 to 200°C. With higher temperature and temperature changes chemical deterioration of detergent chains occur on the surface of nanoparticles, which leads to destabilization of ferrofluid. Ferrofluids durability is e.g., from 8 to 10 years. High quality ferrofluids with long durability are more expensive.

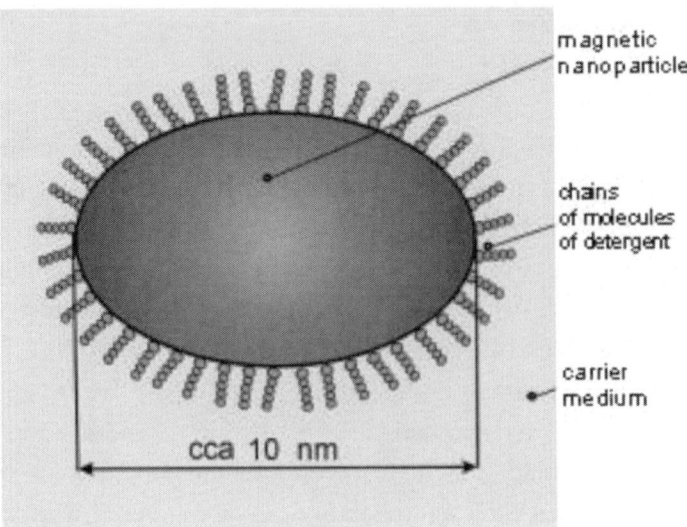

Figure 1. Ferromagnetic nanoparticle with detergent coating in carrier liquid.

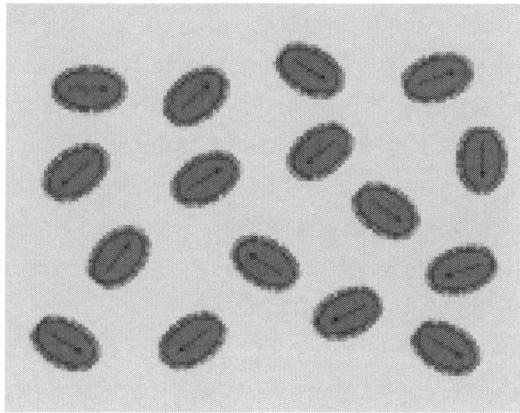

Figure 2. Ferrofluid without the influence of outer magnetic field: magnetic moment of nanoparticles has random distribution.

Nanoparticles move in carrier liquid by thermal (Brown) motion. If the liquid is not in the magnetic field, magnetic moments of nanoparticles are randomly oriented and the liquid is non-magnetic, Figure 2. If the liquid is in the magnetic field, the nanoparticles are polarized, it is they turn in the direction of the magnetic field and make chains lying in the directions of the lines of force. This process leads to considerable changes of physical chemical qualities of ferrofluids. As for their mechanic-elastic qualities it is mainly viscosity – the magnetic viscosity phenomenon. Stable ferrofluid remains liquid even in a strong magnetic field, it means its particles do not sediment and do not aggregate in a strong magnetic field. Unless exposed to the magnetic field, it is isotropic, but in the magnetic field it becomes strongly anisotropic. The dependence of the magnetic inductance on the intensity of the magnetic field has in ferrofluids a course similar to that of solid ferromagnetics: with growing H increases B and asymptomatically approaches the state of saturation. In the linear magnetic field as a result of losses (hysteresis and eddy currents), during demagnetization of nanoparticles these are heated and the carrier liquid is heated as well which leads to decreasing of its viscosity.

In some applications ferrofluids are used with microparticles it is particles of the size from 5 to 15 µm. For these liquids the name is used – *magnetorheoloigical liquids*. Microparticles are multi-domain ones, (non single-domain ones like nanoparticles) and are not magnetically polarized, do not have a magnetic moment. Magnetorheological liquids contain a considerably larger volume of ferromagnetics, up to 70% (of weight). They are not usually stable, it means their microparticles sediment and aggregate, the development of stable magneto-rheological liquids is one of the aims of the present research. They are used in situations when extremely strong magnetic viscosity phenomenon is required, in magnetic field they lose their liquidity and become solid.

Ferrofluids do not exist in nature; they are developed synthetically. Older production technologies were based on long term magnetic crushing of magnetite or ferrite particles in ball mills in detergent solution. The process of grinding lasted from 500 to 1000 hours and after finishing the centrifugal separation of bigger particles followed. At present faster and more effective ways are used, based on various chemical processes leading to precipitation of nanoparticles from solutions of ferrous salts. The obtained product must be purified; it is bigger particles are removed or by centrifugation or by sedimentation caused by gravity or non-homogenous magnetic field.

The producers offer a wide range of ferrofluids or magneto-rheological liquids that differ in their composition, physical chemical qualities and price and they recommend what applications they are suitable. Among important producers of these liquids and equipment using them are e.g., American company Lord Corporation Inc. and British company Liquids Research Ltd.

THE EFFECT OF THE MAGNETIC FIELD: SIMPLE EXPERIMENTS

To understand the behavior of ferrofluids in the magnetic field, some simpler experiments may be presented that show this physically complex environment sometimes behaves contrary to expectations.

Ferrofluid with Free Boundary in Nonhomogenous Magnetic Field

A nonhomogenous magnetic field B affects the fluid. B reaches certain critical value, surface instability of ferrofluids occurs and its surface changes in a system of spikes directed in the course of magnetic lines of force. These spikes are the result of complex structural force ratio in non-linear anisotropic environment of the liquid where magnetic forces apply as well as gravitational force and surface tension. In Figure 3 there is a Petri dish with ferrofluids and permanent magnet underneath Due to the magnetic field, the ferrofluid disintegrates into a number of tiny cones whose axis coincides with the direction of the magnetic field.

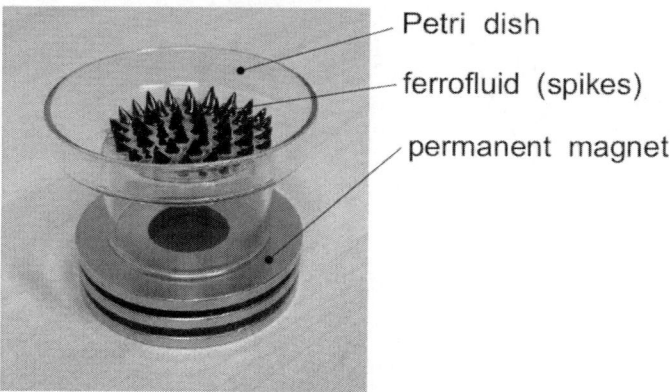

Figure 3. Ferrofluid with free surface under the influence of the magnetic field of a permanent magnet.

Ferrofluid in a Tube

The glass tube contains about 1cm^3 of ferrofluid. Perpendicular to the tube, the magnetic field of the permanent magnets located outside the tube acts. It can be seen from Figure 4 that the ferrofluid is centered in the magnetic field and acts as a plug. This plug can hold a relatively high column of water that is at the top of the tube.

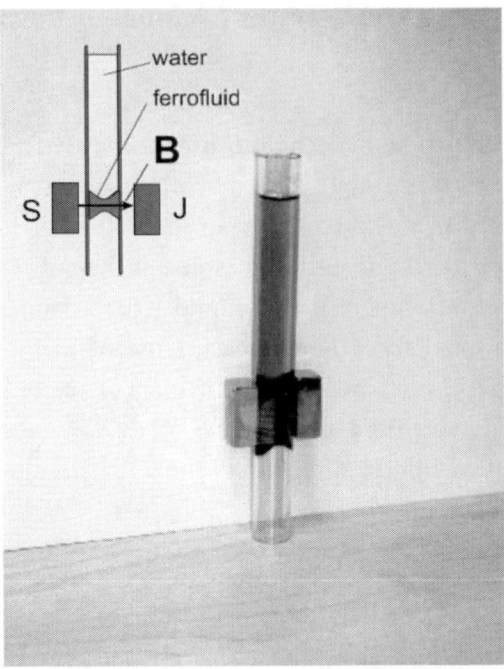

Figure 4. Ferrofluid in a glass tube under the influence of the magnetic field.

Ferrofluid Around a Current-Carrying Wire

In Figure 5 there is a Petri dish with ferrofluid. A conductor which is not yet current-carrying passes through the dish, Figure 5. When current I passes through the conductor it induces, as is known, a magnetic field of intensity in the vicinity of the conductor $H = \dfrac{I}{2\pi r}$, where r is the perpendicular distance from the conductor. Ferrofluid is drawn into this inhomogeneous magnetic field, so that its originally horizontal level (according to Figure 5) significantly changes its shape as shown in Figure 6. In the axial cross-section of the dish, the level of ferrofluid bonded is bonded by the curve $y = \text{const.}/r.$, equinox hyperbola.

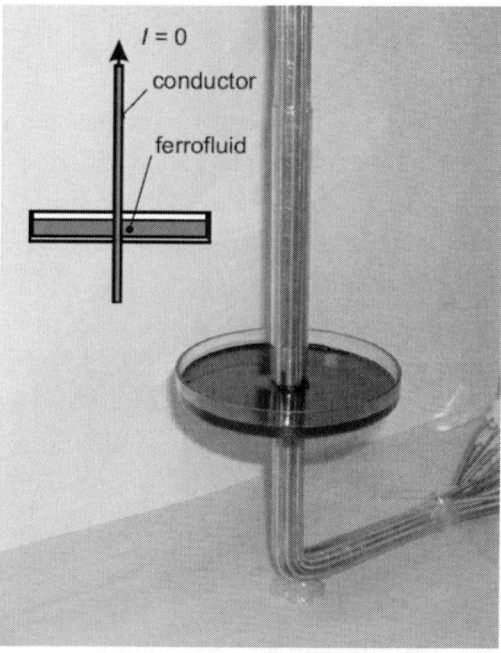

Figure 5. Petri dish with ferrofluid without electric current.

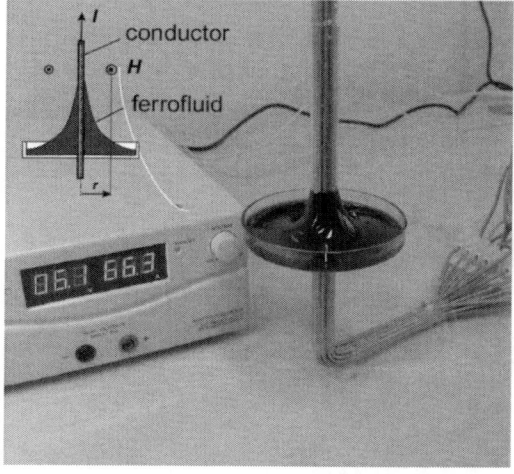

Figure 6. Same arrangement a Figure 5, but current I = 66,3 A.

LEVITATION IN A FERROFLUID

The following observation applies to a permanent magnet (Samuel Earnshaw, 1839): A permanent magnet cannot be stably levitated by a stationary magnetic field. This means, for example, that a passive magnetic bearing cannot be built without thrust bearings that keep the shaft in a stable position. A completely different situation arises when the medium in which levitation takes place is ferrofluid. If you place a sealed container of ferrofluid in an inhomogeneous magnetostatic field, the ferrofluid will be drawn into places of increasing field strength. It does not appear externally, but the stress in the ferrofluid is present, Figure 7. Inserting a nonmagnetic body into the ferrofluid now breaks the internal stress in the liquid and causes forces to act on the body (together with gravitational force, Figure 8). This phenomenon is called *passive levitation of a nonmagnetic body* [17, 18].

Levitation in the ferrofluid can also be monitored in other ways. When a permanent magnet is immersed in a ferrofluid, it is subjected to a magnetic force which, together with gravity, makes it possible to achieve the equilibrium position of the magnet in which it levitates, Figure 9. This phenomenon is called the *permanent magnet levitation*.

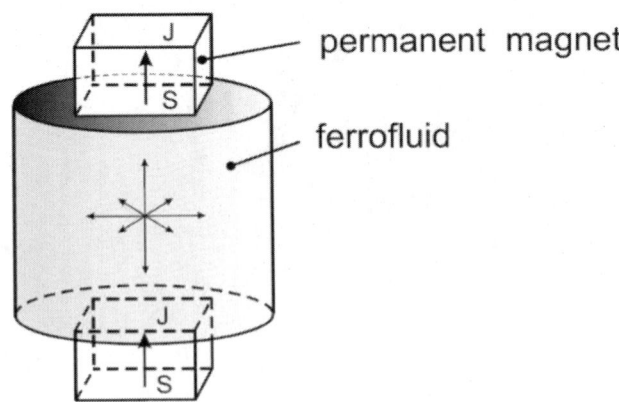

Figure 7. Ferrofluid under the effect of external magnetic field.

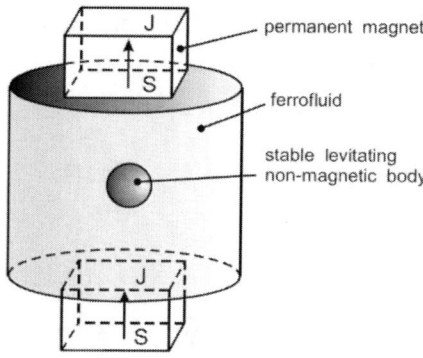

Figure 8. Ferrofluid under the external magnetic field – non-magnetic body levitates.

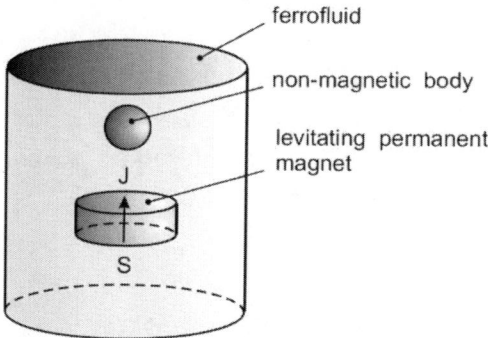

Figure 9. Ferrofluid with levitating permanent magnet.

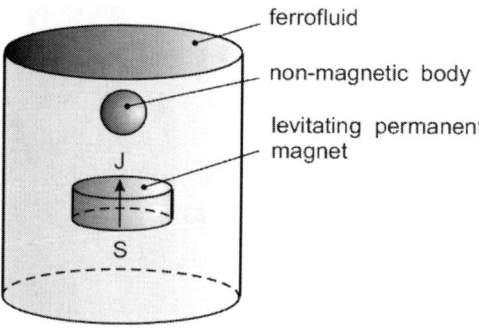

Figure 10. Ferrofluid with levitating non-magnetic body and permanent magnet.

A modification of these phenomena is represented by a permanent magnet immersed in a magnetic liquid in the presence of a non-magnetic body, Figure 10. Both the non-magnetic body and the permanent magnet are levitated.

MATHEMATICAL DESCRIPTION OF FERROFLUIDS AS ANISOTROPIC MEDIA

Ferrofluids exhibit material (microscopic) anisotropy, the essence of which is based on the physical structure of ferrofluids discussed in the previous chapter. The material characteristics expressing the electromagnetic properties of the isotropic medium, such as permittivity ε, conductivity γ and permeability μ, are scalars. In anisotropic environments, i.e., ferrofluids, these characteristics are more complex mathematical quantities: they are tensors [13, 14].

MATHEMATICAL DESCRIPTION OF ANISOTROPY

Anisotropy will be illustrated based on the mechanical analogy. Figure 11 shows two identical springs which lie in the direction of the orthogonal coordinate axes (x, y). Both springs are connected together in the middle (i.e., at the beginning) and are fixed at the ends. If we attach a string to the point of connection of the springs and pull it so that it moves from the beginning to point A(x, y), the direction of deflection of the joint r is identical with the direction of the return force F (Figure 12). This is evident from this consideration. For small deflections, the x-deflection (i.e., the x-coordinate of point A) is caused by the x-component of the force F_x, and the y-coordinate is likewise:

$$F_x = -k\,x \quad \text{a} \quad F_y = -k\,y \tag{1}$$

where k is the constant of Hook's law. Since both springs are the same, this constant is also the same for both force components. So it is

$$\mathbf{F} = \mathbf{i}\,F_x + \mathbf{j}\,F_y = -k\,(\mathbf{i}\,x + \mathbf{j}\,y) = -k\,\mathbf{r} \qquad (2)$$

thus the vectors $\mathbf{F}$ and $\mathbf{r}$ have the same direction. The system has the same mechanical properties in the x and y directions, and is therefore isotropic. It can be characterized by a single scalar constant k.

Let the two springs be different. Consequently, their properties are described by constants $k_x \neq k_y$ (Figure 12). Equation above then passes to

$$\mathbf{F} = -\mathbf{i}\,k_x\,x - \mathbf{j}\,k_y\,y \qquad (3)$$

This can be written as matrix:

$$\mathbf{F} = -\mathbf{K}\,\mathbf{r} \qquad (4)$$

where

$$\mathbf{F} = \begin{bmatrix} F_x \\ F_y \end{bmatrix}, \quad \mathbf{K} = \begin{bmatrix} k_x & 0 \\ 0 & k_y \end{bmatrix}, \quad \mathbf{r} = \begin{bmatrix} x \\ y \end{bmatrix} \qquad (5)$$

The quantity $\mathbf{K}$ is a *tensor* which can be expressed by a square matrix. The mechanical system considered has different properties in the x direction than in the y direction, so it is an *anisotropic* system. It is characterized by two scalar variables (k_x, k_y), respectively by diagonal matrix $\mathbf{K}$, which is second order.

If we introduce a rotated coordinate system (x', y'), it is found that equation mentioned is valid too, but the tensor $\mathbf{K}$ will be expressed by a full matrix, i.e., the system will be characterized by four scalar quantities. It can be shown that in every anisotropic environment a coordinate system (x, y) can be introduced so that the tensor matrix $\mathbf{K}$ is diagonal. The x and

y axes are then called the major axes. Anisotropic media with established major axes are sometimes referred to in the literature as orthotropic media.

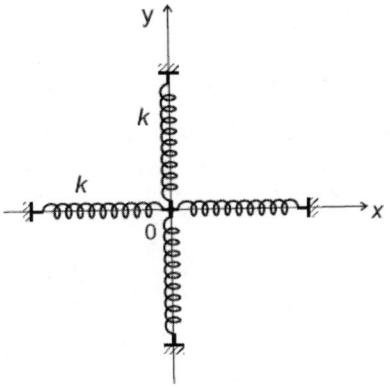

Figure 11. Mechanical model of isotropic system.

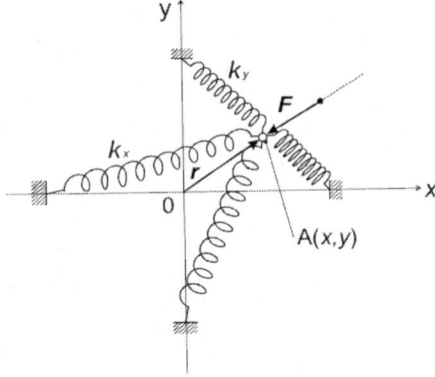

Figure 12. Mechanical model of anisotropic system.

The mechanical system considered would probably be designed as spatial. It would be true for him again.

$$\boldsymbol{F} = \begin{bmatrix} F_x \\ F_y \\ F_z \end{bmatrix}, \quad \mathbf{K} = \begin{bmatrix} k_{11} & k_{12} & k_{13} \\ k_{21} & k_{22} & k_{23} \\ k_{31} & k_{32} & k_{33} \end{bmatrix}, \quad \boldsymbol{r} = \begin{bmatrix} x \\ y \\ z \end{bmatrix} \qquad (6)$$

If we choose the coordinate system (x, y, z) in the main axes, the **K** is represented by a diagonal matrix. We can therefore imagine this tensor as a quantity that converts the vector $\boldsymbol{F}$ to another vector $\boldsymbol{r}$, which generally differs in all its characteristics: size, direction, sense and units. Under certain circumstances, the tensor may become scalar (in the case of Figure 12, when $k_x = k_y$).

MATERIAL EQUATIONS FOR FERROFLUID AS ANISOTROPIC MEDIUM

In an *isotropic* environment, the material equations of the electromagnetic field are:

$$\boldsymbol{D} = \varepsilon \boldsymbol{E}, \quad \boldsymbol{J} = \gamma \boldsymbol{E}, \quad \boldsymbol{B} = \mu \boldsymbol{H} \qquad (7)$$

where the material characteristics ε, γ, μ are scalars. In *anisotropic environment*, similar relationships apply, but instead of scalars ε, γ, μ material characteristics are second order tensors. We denote them as $\boldsymbol{\varepsilon}$, $\boldsymbol{\gamma}$, $\boldsymbol{\mu}$. For example, for electrical induction

$$\begin{aligned} \boldsymbol{D} &= \boldsymbol{\varepsilon}\, \boldsymbol{E}, \\ D_x &= \varepsilon_{xx} E_x + \varepsilon_{xy} E_y + \varepsilon_{xz} z E_z \\ D_y &= \varepsilon_{yx} E_x + \varepsilon_{yy} E_y + \varepsilon_{yz} E_z \\ D_z &= \varepsilon_{zx} E_x + \varepsilon_{zy} E_y + \varepsilon_{zz} E_z \end{aligned} \qquad (8)$$

This set of equations can also be written in the form

$$D_k = \sum_{l=1}^{3} \varepsilon_{kl} E_l \qquad (9)$$

where ε_{kl} are components of the *permittivity tensor*, which can be described with matrix

$$\boldsymbol{\varepsilon} = \begin{bmatrix} \varepsilon_{11} & \varepsilon_{12} & \varepsilon_{13} \\ \varepsilon_{21} & \varepsilon_{22} & \varepsilon_{23} \\ \varepsilon_{31} & \varepsilon_{32} & \varepsilon_{33} \end{bmatrix} \qquad (10)$$

Its diagonal components ε_{kk} are called *normal permittivities* and nondiagonal components ε_{kl} ($k \neq l$) are *tangential permittivities*. It can be shown that the permittivity tensor $\boldsymbol{\varepsilon}$ is symmetric, i.e., $\varepsilon_{kl} = \varepsilon_{lk}$, i.e., the anisotropic dielectric is generally determined by six scalars, which are constants when the dielectric is linear, or by functions of vector $\boldsymbol{E}$ if the dielectric is nonlinear, or function of coordinates if the dielectric is inhomogeneous.

The symmetry of tensor $\boldsymbol{\varepsilon}$ allows to select such a coordinate system (x, y, z), where axes lie in the principal axes of the dielectric. In this coordinate system, the permittivity tensor matrix is diagonal and the expressions above change to

$$D_x = \varepsilon_{xx} E_x, \quad D_y = \varepsilon_{yy} E_y, \quad D_z = \varepsilon_{zz} E_z \qquad (11)$$

It can be seen from listed equations that if we insert an anisotropic dielectric into an electric field whose vector $\boldsymbol{E}$ is not parallel to any of the major axes, the vectors $\boldsymbol{E}$ and $\boldsymbol{D}$ are not parallel (collinear), Figure 13. On the other hand, from the matrix it follows that if the direction of vector

E coincides with any of the major axes, the vectors E and D are collinear, Figure 13 Figure 14 shows the field lines and induction lines of the electric field of a plate capacitor with an anisotropic dielectric whose principal axes (x, y) are not parallel (or perpendicular) to the dielectric surface.

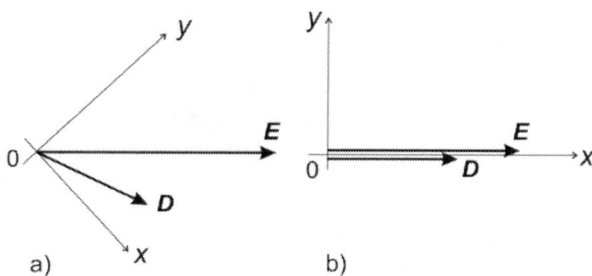

Figure 13. Vectors E and D in anisotropic dielectric, if a) both vectors are in general position with respect to the major axes (x, y), b) vector E has the same direction as the major axis.

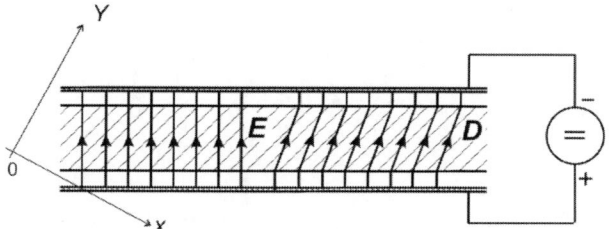

Figure 14. Vectors E and D in anisotropic dielectric of plate capacitor at a general position related to the major axes (x, y).

Similar relationships as for electric field apply to current field in anisotropic conductive medium, characterized by *conductivity tensor* γ and to magnetic field in anisotropic magnetics, characterized *by permeability tensor* μ.

EQUATIONS OF STATIONARY ELECTRIC AND MAGNETIC FIELD FOR FERROFLUIDS AS ANISOTROPIC ENVIRONMENT

Electromagnetic field is usually solved as a boundary value problem for scalar potential φ or for vector potential $\mathbf{A}$. This concept can be used also for solving electromagnetic field in anisotropic environment, that is in a region that is partially or completely filled with a ferrofluid, but the equations for potentials will have different form than for isotropic environment. As an example, consider a stationary current field in an anisotropic conductive environment and formulate an equation for a scalar potential. We choose a system of coordinate axes (x, y) so that it coincides with the principal axes of the conductivity tensor $\mathbf{G}$. From the material equation and from the scalar potential definition it follows

$$\mathbf{J} = \mathbf{G}\mathbf{E} = -\mathbf{G}\,\text{grad}\,\varphi = \tag{12}$$

$$= -\mathbf{i}\gamma_x \frac{\partial \varphi}{\partial x} - \mathbf{j}\gamma_y \frac{\partial \varphi}{\partial y} - \mathbf{k}\gamma_z \frac{\partial \varphi}{\partial z}$$

Using the continuity equation $\text{div}\,\mathbf{J} = 0$,

$$\frac{\partial J_x}{\partial x} + \frac{\partial J_x}{\partial y} + \frac{\partial J_x}{\partial z} = 0 \tag{13}$$

We obtain the *basic equation* for calculating the electric potential distribution of an electric field in an anisotropic environment:

$$\gamma_x \frac{\partial \varphi^2}{\partial x^2} + \gamma_y \frac{\partial \varphi^2}{\partial y^2} + \gamma_z \frac{\partial \varphi^2}{\partial z^2} = 0 \tag{14}$$

From this eqation it is apparent that in the anisotropic conductive medium the Laplace equation no longer applies, but the equation is somewhat more general. By transforming coordinates (called isotropic transformation), this equation can be converted to Laplace's equation (boundary conditions must also be transformed) and then solved by conventional methods. In practice, numerical methods are particularly suitable.

A similar procedure and equation applies to the scalar potential of an electrostatic field and a magnetostatic field. Equations for vector magnetic potential A can be formulated in a similar way.

FERROHYDRODYNAMICS THEORY

In this part we introduced a short synopsis of mathematical description of system with the ferrofluid, in the form of boundary value coupled problem, based on magnetically/mechanical fields. Magnetic field in the domain Ω is described by the equation

$$\mathrm{rot}\frac{1}{\mu}\mathrm{rot}\,A(r) = -J, \quad A \in \Omega \tag{15}$$

together with boundary condition for A, where by vector r is defined the point in the field, μ is the permeability, γ is the conductivity and J is the current density. The magnetic field strength is

$$H = \frac{1}{\mu}\mathrm{rot}\,A \qquad (\mathrm{div}\,A = 0) \tag{16}$$

For 2D rotational symmetric system (r, z) with ferrofluid (see e.g., Figure 4) hold true

$$\frac{\partial}{\partial z}\left(\frac{1}{\mu}\frac{\partial A_\varphi}{\partial z}\right) + \frac{\partial}{\partial r}\left(\frac{1}{\mu r}\frac{\partial}{\partial r}(r A_\varphi)\right) = -\mathbf{J} \qquad (17)$$

$A_\varphi \in \Omega$, together with boundary condition and the magnetic field strength is

$$H_r = -\frac{1}{\mu}\frac{\partial A_\varphi}{\partial z}, \quad H_z = -\frac{1}{\mu}\left(\frac{\partial A_\varphi}{\partial r} + \frac{A_\varphi}{r}\right) \qquad (18)$$

From generalized Navier-Stokes equations of conventional fluid mechanics may be deduced for incompressible isotropic ferrofluids and for the motionless system the ferrodynamic Bernoulli equation. In the steady state has the form [15]:

$$p^* + \rho g h - \mu_0 \overline{M} H = \text{const.}, \quad p \in \Omega \qquad (19)$$

with following boundary condition:

$$p^* + p_n = p + p_c \qquad (20)$$

where $p^* = p + p_s + p_m$ is composite pressure, p is thermodynamic pressure, p_s is magnetostrictive pressure, $p_m = \mu_0 \overline{M} H$ is fluid-magnetic pressure, where $\overline{M} = \int_0^H M \, dH$, $p_n = \frac{1}{2}\mu_0 M_n^2$ is magnetic normal traction, p_c is capillary pressure, p_0 is pressure in nonmagnetic fluid, ρ is particle mass density and $g = 9{,}8$ m/s.

For some applications the *dynamics of magnetic viscosity phenomenon* is important (e.g., for ferrohydrodynamic damper). In this cases are important the determination of the response of viscosity to the change of the outer magnetic field. Its value is calculated in nanoseconds.

MEASUREMENT OF PERMEABILITY/PERMITTIVITY TENSOR OF FERROFLUIDS

The magnetic field acting on the ferrofluids causes microstructural conversions that result in a change of permeability ferrofluids [16]. For this physical phenomenon introduce labelling field induced magnetic. For ferrofluids which are in the state FIMA is described the experimental method for the examination of their permeability tensor. It is also described analogous phenomenon in where there is a change of permittivity ferrofluids. We call it field induced dielectric anisotropy (FIDA). The contribution describes the method of measuring of the permittivity tensor. It can be expected that the FIMA and FIDA of ferrofluids will find interesting applications in designing of various sensors, in measurement technology, in mechatronic and in other areas of practice.

A dominant feature of ferrofluids is the magnetoviscous effect [8]. In engineering applications it is widely used, namely for controlled ferrohydrodynamic damper, ferrohydrodynamic sealing etc. (see e.g., [19, 22]). Ferrofluids, however, have other interesting properties, some of which still have not been explored sufficiently. In the present study we will examine the effect of an external magnetic field on the permeability and permittivity of ferrofluids. On the basis of experimental methods to demonstrate that these dependencies are greatly significant under the effect of a relatively weak external magnetic field. The authors of this work are convinced that these findings will be sources of various interesting applications.

INDUCED ANISOTROPY

Microstructure of the ferrofluids consists of a chain of dipolar nanoparticles, together with a surfactant, in a carrier liquid. In the absence of an external magnetic field, these chains are arranged randomly. External magnetic field causes the rotation of the chains into direction of the magnetic field (Figure 15). This conversion of the microstructure of ferrofluids results in a change of its viscosity and can expect further changes of its physical properties. The question is whether these changes will be quantitatively significant. In the next phenomenological study, we investigate the changes permeability and permittivity.

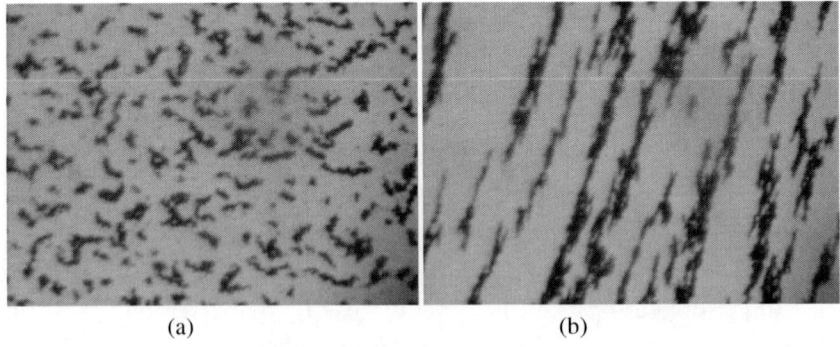

(a) (b)

Figure 15. Microstructure of ferrofluids: a) without external magnetic field, b) with an external uniform magnetic field.

Direct observation of the microstructure of ferrofluids by electron microscopy, when simultaneously acting external magnetic field, is very difficult. The problem lies in the need to prevent interaction between the observed object (i.e., dipolar magnetic nanoparticles) including the external magnetic field and the magnetic lens of the electron microscope. For this purpose an electroncryo-microscope was used in [12] and the observed object was shaped as a thin layer. Microstructure observations of ferrofluids were not done because this information is not crucial for the studied phenomena.

FIELD INDUCED MAGNETIC ANISOTROPY

From the above it is clear that without application of external magnetic field the ferrofluid is (from macroscopic viewpoint) an isotropic medium (Figure 15a), while after application of the external magnetic field becomes magnetically anisotropic medium (Figure 15b). The physical phenomenon where the action of an external magnetic field leads to structural changes of ferrofluids and it becomes magnetically anisotropic medium, is called field induced magnetic anisotropy (FIMA). Magnetic anisotropy thus occurs as a response of ferrofluids to the external magnetic field. FIMA depends on the intensity of external magnetic fields. In the maximal stage of FIMA occur in such an external magnetic field, in which perfectly ordered chains arise.

The magnetic field distribution in such as to minimise its energy. The easy axis of anisotropic medium, in the state of FIMA, are curvilinear and identical the vector lines of magnetic field B. The direction of vector B determines the local orientation of the easy axis. Only in a magnetic field whose vector B has only one component, $B = i\ B_x$ (for example in a homogeneous magnetic field) the anisotropy has a rectilinear easy axis. That the case of oriented iron sheets. If one neglect hysteresis of ferrofluids, then if the external magnetic field does no act, FIMA will disappear and the ferrofluids become isotropic again, including all of their original physical properties. Thus FIMA is a reversible phenomenon.

The magnetic properties of ferrofluids in a state of FIMA are characterized by a permeability tensor. For 2D magnetic field this tensor is expressed by diagonal matrix

$$\mu(x,y) = \mu_0 \begin{bmatrix} \mu_{rx} & 0 \\ 0 & \mu_{ry} \end{bmatrix}$$

(21)

where μ_0 is the permeability of free space, μ_{rx} and μ_{ry} are the permeability in x and in y direction. The easy axis x of anisotropy has a tangent-direction to the field line of external magnetic field and *y*-axis in that point is perpendicular to it.

Field Induced Dielectric Anisotropy

Microstructural changes in ferrofluid caused by an external magnetic field, may also affect other their physical properties. The changes of permittivity are called field induced dielectric anisotropy (FIDA). Dielectric properties of ferrofluids are then characterized by *permittivity tensor*, which for 2D the electric field has the form

$$\varepsilon(x,y) = \varepsilon_0 \begin{bmatrix} \varepsilon_{rx} & 0 \\ 0 & \varepsilon_{ry} \end{bmatrix}$$

(22)

where ε_0 is the permittivity of free space and ε_{rx} and ε_{ry} are the relative permittivity in *x*- and in *y*- direction. Additional properties of the FIDA are analogous as those of the FIMA.

"Frozen" Anisotropy

If the ferrofluids with induced anisotropy changes its state (e.g., it will "freeze") FIMA and FIDA become permanent effects, even if the external magnetic field disappears. The induced magnetic anisotropy is an irreversible phenomenon. The change of state ("freezing") of ferrofluids may occur not only by changing of the temperature, but also by chemical reaction such as when the carrier medium of ferrofluids comprises a monomer which is activated by treatment with an initiator of ferrofluids at high polymer solids. This technology could be used to create magnetic circuits in solid state, with a perfect orientation according to magnetic field of an excitation coil. In the design of magnetic circuits would not be necessary to carry out the difficult implementation of cuts and composing the strips of the magnetic circuit from oriented sheets. Unfortunately, the current magnetic fluids have very low permeability, hence the production of such oriented magnetic circuit has no practical significance.

Solid medium in the state induced anisotropy has extremal characteristics, e.g., magnetic circuits have a minimal reluctance. Similarly behave devices with ferrofluids using an electric field. Capacitors with the dielectric formed as a "frozen" ferrofluids can reach maximum capacitance.

INVESTIGATIONS OF PERMEABILITY TENSOR

Components of permeability tensor μ_{rx} and μ_{ry} will be determined by using self-inductance coil test. For this purpose, the test coil is placed in a closed housing which has the shape of a cube. The housing may be filled with the measured ferrofluids. An excitation coil is placed on the housing and its magnetic field brings the ferrofluid into the FIMA state. The housing including the exciting coil is then inserted into the magnetic circuit from the highly permeable material. The described device is shown in Figure 16. The test coil in the housing has either transverse (Figure 16a) or longitudinal (Figure 16b) position. With this configuration it is achieved that in the area of the test coil the excitation coil induces a magnetic field necessary for the creation FIMA, which is homogeneous. Also, the magnetic field which induces test coil during measuring the self-inductance of coil will be homogeneous. Measuring, proceed as follows:

Housing is not filled with ferrofluids. In the surroundings of the test coil is therefore air and the self-inductance of the coil is

$$L_{0i} = \frac{N^2}{R_m + R_{mi} + R_{mc}}, \quad i = x, y \tag{23}$$

where N is the number of turns of the test coil, R_m, R_{mx}, R_{my}, R_{mc} are the magnetic reluctances of the sections of the part, which closes the magnetic reluctances field of test coil. These have lengths l, δ_x, δ_y, l_m. The magnetic shielding circuit has a high permeability and thus its reluctance is $R_{mc} \rightarrow 0$. Since the magnetic field of test coil is uniform, determination of reluctances R_m, R_{mx}, R_{my} are easy. The inductance of the test coil in the transverse position ($i = x$), or in the longitudinal position ($i = y$) is then

$$L_{0i} = N^2 \frac{\mu_0 s}{l + 2\delta_i}, \qquad i = x, y \tag{24}$$

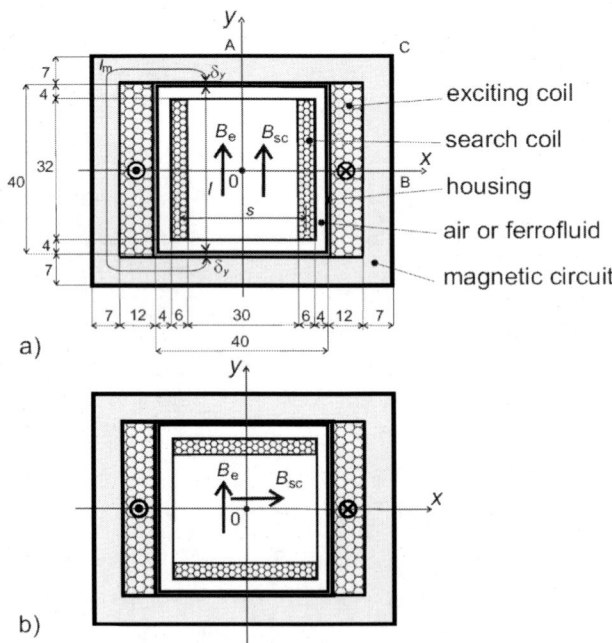

Figure 16. The device for measuring the permeability tensor components μ_{rx}, μ_{ry} of ferrofluids; a) Test coil in longitudinal position, b) Test coil in the transverse position.

Housing will be filled with ferrofluids and the excitation coil is powered by direct current; the ferrofluid will be in the state of anisotropy. Set such excitation current in order to saturated magnetic polarization of ferrofluids. Inductance test coils in the transverse and longitudinal position is then

$$L_i = N^2 \frac{\mu_0 \mu_{ri} s}{l + 2\mu_{ri} \delta_i}, \qquad i = x, y \qquad (25)$$

We measure inductance of coil test L_{0x}, L_{0y}, L_x, L_y and from the equations above we calculate the of permeability tensor:

$$\mu_{ri} = \frac{L_i l}{L_{0il} + 2\delta_i (L_{0i} - L_i)}, \qquad i = x, y \qquad (26)$$

COMPUTER SIMULATION OF FIELD INDUCED MAGNETIC ANISOTROPY

The described method of examination of the permeability tensor of the ferrofluid which is in the state of FIMA, is based on the assumption that the excitation coil induces a magnetic field that is strong enough to achieve a saturated magnetic polarization of the ferrofluids, i.e., "perfect" arrangement of the chains of nanoparticles and that this field is homogeneous. Furthermore, it is required that the magnetic field induced by the test coil for measuring the inductance be also homogenous. Deviations from these assumptions must lie within acceptable limits. Verification of these assumptions is performs with computer simulations.

Both investigated magnetic fields (i.e., the magnetic field of excitation coils and magnetic field of test coils) are 3D. To verify the above properties we will analyze them as a 2D field. Solution of the magnetic field in the plane (x, y) and in the z direction (direction perpendicular to the plane of the drawing) counts as "long" configuration. For the reasons of symmetry, it is sufficient if the definition area for the field will be in the first quadrant of the plane (x, y), Figure 17.

Numerical solution of the magnetic field was carried out by finite element method (FEM) with program Quick-Field 5.0. The following pictures show the magnetic field lines of excitation coils (Figure 18a), wherein the test coils in the transverse position (Figure 18b) and then in longitudinal position (Figure 18c). These results are valid for relative permeability $\mu_r = 1$, for higher values μ_r (e.g., for $\mu_r = 5$) the shape of the magnetic field similar. From Figure 18abc it is obvious that the assumption of homogeneity of the magnetic field in the vicinity the measuring coil is achieved with sufficient accuracy. The homogeneity of the magnetic field was confirmed more precisely by numerical calculating the values of the components B_x and B_y in different positions of the plane which is parallel to plane (x, z), resp. (y, z).

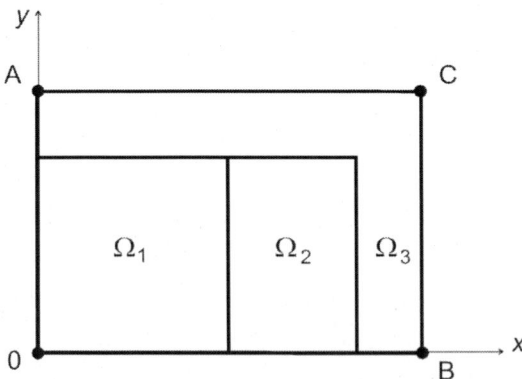

Figure 17. Definition area. Subareas: Ω_1 ... operating room, Ω_2 ... excitation coil, Ω_3 ... magnetic circuit.

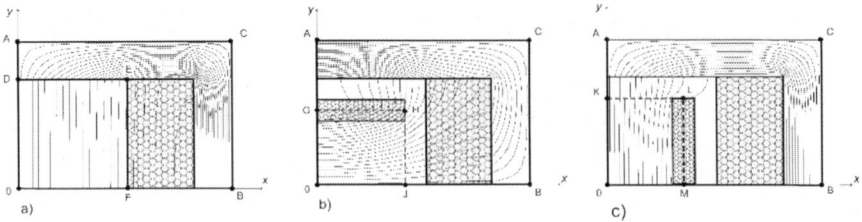

Figure 18. Magnetic field lines: a) in the operating area of 0-F-E-D, b) in the area of the test coil in the transverse position 0-J-H-G, c) in the area of the test coil in the longitudinal position 0-M-L-K.

INVESTIGATIONS THE PERMITTIVITY TENSOR

For measuring the permittivity of ferrofluid the capacitance was used of the capacitor test, the dielectric of which is both air and a ferrofluid, Figure 19. First we measure the capacitance C_0 of the capacitor, the dielectric of which is air. Then, after filling the housing with ferrofluid and inserting it in a magnetic field. Then ferrofluid is in the state of FIDA. The magnetic field in the entire volume of the housing has only component B_x and is homogeneous. We measure the capacity C_x. Then we set the housing so that the magnetic field has only component B_y; we measure the capacitance C_y. Permittivity tensor components may be calculated from the simple relationship

$$\varepsilon_{ri} = \frac{C_i}{C_0}, \quad (i = x, y) \tag{27}$$

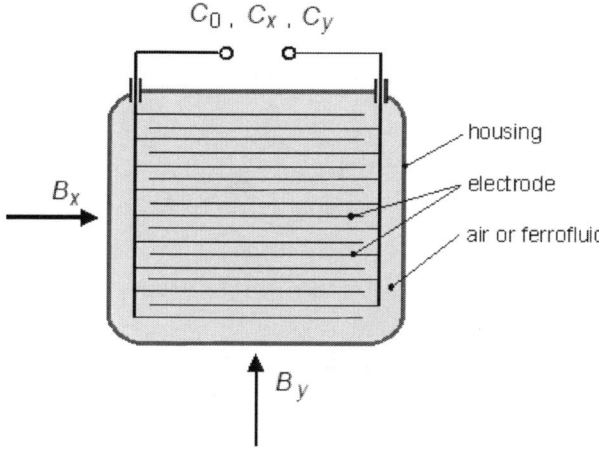

Figure 19. A device for measuring components ε_x, ε_y of permittivity tensor of ferrofluids.

EXPERIMENTS

The aim of this chapter is to draw attention to the interesting property of ferrofluids going from the isotropic to anisotropic state, and vice versa. This phenomenon was called FIMA, or FIDA and could find practical application in various fields of technology, e.g., in measuring technology, in sensors, microelectromechanical systems, etc. A method has been proposed for measuring the permeability tensor, or, permittivity tensor. This method has been implemented for the ferrofluids for the ferrofluids from different manufacturers. For example, when Ferro-fluids Ferrotec (USA), catalogue No. EFH 1 3-200002 were in the state of FIMA (with saturated polarization) the values of the components of relative permeability changed by about 34%. When the same ferrofluids were in a state of FIDA the measured changes of relative permittivity were around around 24%. The authors believes that it is meaningless to indicate all measured values, as for other ferrofluids the values and μ and ε are substantially different. It is expected that it will be possible to choose a ferrofluid, in which the changes of μ and ε are much more pronounced. However, the measured inductance, or capacitance, and thus the identified permeability μ_{rx}, μ_{rx} and permittivity ε_{rx}, ε_{ry}, depend on the degree of arrangement of dipolar nanoparticles investigated ferrofluids, namely on the value of the excitation of magnetic field. The induced anisotropy reaches the maximum value in an external magnetic field, in which the saturated magnetic polarization of ferrofluids occurs. It is important that for a certain class of ferrofluids the observed phenomena are considerably significant and therefore usable in applications.

Interesting could be research of induced dynamic anisotropy, which arises when an external time-variable magnetic field acts on the ferrofluids.

MAGNETIC FIELD CONTROLLED CAPACITOR

A capacitor using a magnetic fluid as a dielectric is presented in this paper. The capacity of this capacitor can be controlled with an external magnetic field.

Without the effect of an external magnetic field, ferromagnetic particles are oriented randomly in the liquid. The external magnetic field causes the particles to group into chains. These chains copy the course of the magnetic field lines of force. These structural changes cause that the magnetic fluid ceases to behave as an isotropic medium and becomes an anisotropic material, this effect is called *magnetic induced anisotropy*. Macroscopically, physical properties of the liquids change. The dominant property of magnetorheological fluids is their viscosity change, the *magnetorheological effect*, which is used in several technical applications, e.g., magnetic field controlled dampers, or crankshaft seals [19–22].

THE PHYSICOCHEMICAL NATURE OF THE MAGNETIC FIELD CONTROLLED CAPACITOR

Without an external magnetic field affecting the fluid, the ferromagnetic particles are randomly organized and the magnetic fluid behaves as an isotropic dielectric, which can be characterized by a scalar variable - the permittivity $\varepsilon = \varepsilon_o\, \varepsilon_r$, where $\varepsilon_o = 8{,}85 \cdot 10^{-12}$ Fm^{-1} and ε_r is the relative permittivity.

The external magnetic field causes structural changes and the fluid becomes anisotropic (Figure 15) and when properly established cartesian (Figure 20), can be characterized with the permittivity tensor $\boldsymbol{\varepsilon}$ (see equations in previous chapter). This effect will be called *magnetic field induced dielectric anisotropy* – MFIDA.

While without the external magnetic field effect is the capacity of the capacitor $C_0 = C(\varepsilon_r)$, with the effect of the external field B_y is its capacity $C = C(\varepsilon_{ry}) \neq C_0$. The magnetic induction of the external field B_y determines the particles chaining rate (till the saturation) and with it the magnitude of the permittivity tensor components ε_{rx}, ε_{ry}, therefore the capacity of device is $C = C\{\varepsilon_{ry}(B_r)\}$.

The arrangement of ferromagnetic particles, therefore the value of capacity C, is generally dependant on the intensity of the electric field E_y as well. Until now, no voltage U on the capacitor electrodes was considered. If $U \neq 0$, the arrangement of the fluid structure (and its permittivity) is affected by the intensity E_y as well, the capacitor with liquids dielectric is nonlinear consequently. This effect will be called *electric field induced dielectric anisotropy* – EFIDA. Both of these phenomena, MFIDA and EFIDA, act simultaneously. In experiments executed by us, the effect of EFIDA was negligible in experiments executed by us, therefore $C(U) \sim$ const.

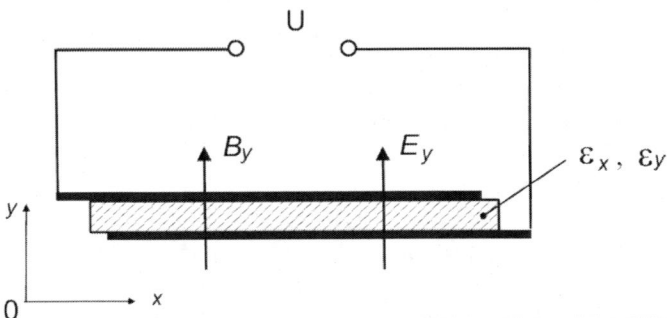

Figure 20. Magnetic field controlled capacitor cartesian establishment.

Due to the ferromagnetic particles weight and the viscosity of the carrier liquid, mechanical dynamics applies when ferromagnetic particles chains are created and the chaining shows certain time delay. This is apparent from the time sequence of photomicrographs showed in Figure 21.

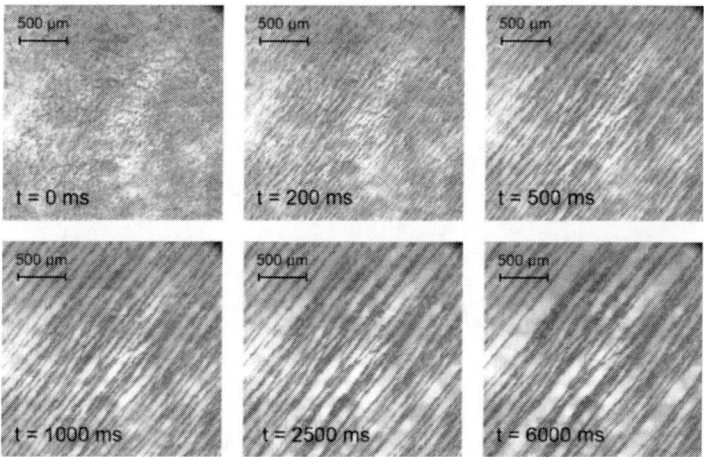

Figure 21. High speed camera photomicrograph, the time delay of the MRHCCS-4B fluid ferromagnetic particles chaining is apparent. The external field $B_y = 22$ mT was applied on the fluid at the time $t = 0$.

DESIGN OF EXPERIMENTAL SAMPLES

An experimental multilayer planar capacitor was made with the use of a 3D printer [31]. The electrodes are made from an aluminium sheet and linked together with copper conductors. The area between electrodes was filled with the magnetic liquid MRHCCS4-B provided by the Liquids research, Ltd. The thickness of the layers was chosen in such a way that no mechanical sag of electrodes causing the change of capacity can occur. The housing and non conductive parts of the device were printed from plastic (concretely ABS) and two supplies are lead out of the capacitor. To prevent incidental leak of the fluid, the outside of the device was cemented and varnished.

EXPERIMENTS

A common alternate bridge was used to measure the capacity of the controlled capacitor prototype. To verify the measured values, the capacity of the device was determined from the time response to its charging as well to eliminate appropriate dynamic changes caused by the current powering the bridge. The time of this response was intentionally prolonged by a serial resistance. Results obtained by both methods were closely comparable.

$C = F(B)$ CHARACTERISTICS

The capacitor prototype was placed together with the Hall probe of teslameter Elimag MP-1 between the poles of (previously demagnetised) magnetic circuit and connected to the RLC measuring bridge U1733C. Gradually increasing the powering current to the magnetic circuit, external magnetic field affecting the capacitor B_y rises, values of B_y and C were registered. To examine the dynamics of the micro structural changes of the used fluid, the magnetic field was gradually incremented with a constant time step between individual increments. To investigate the hysteresis, the magnetic field was gradually lowered after, again with a constant time step. Measured characteristics can be seen in Figure 22.

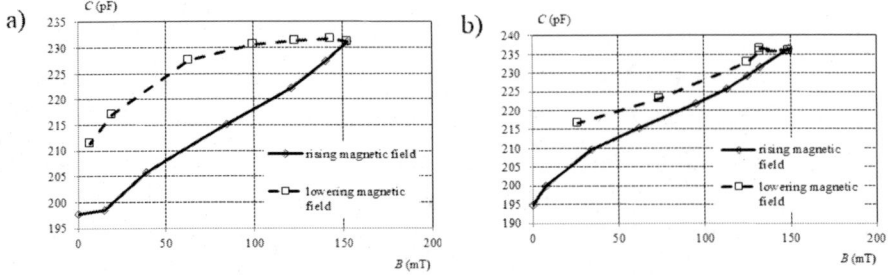

Figure 22. Hysteresis of the dependence $C(B)$ of the measured capacitor with magnetic fluid. The time interval between individual measurements was $\Delta T = 30$ s (Figure a) and $\Delta T = 60$ s (Figure b).

Time to reach the steady state of the capacity of the measured capacitor C, after removing the external magnetic field, is in order of minutes.

LEAKAGE CURRENT

No leakage current was measured when the capacitor was long-term connected to a DC voltage with value $U_0 = 100$ V.

MEASURING OF EFIA, I.E., $C = F(U)$ CHARACTERISTICS

The characteristics of $C = f(U)$ was constructed from the gradual charging of capacitor after it was connected to a DC voltage source. The voltage U_0 was jump in individual steps and generated transient was measured with an oscilloscope. To slower the time response, a resistor $R = 33$ kΩ was serially connected to the measured capacitor. From the relation for the time constant $\tau = RC$, values of the capacity were determined after, while the time to reach the value of $U_C = 0{,}63\ U_0$ was taken for the time τ. When changing the voltage U_0 in the interval <0.5; 4 V>, the capacity with deviations $C = 198 \pm 5$ pF with random distribution was measured. These deviations were considered to be the error of the used method for determining the capacity, not for the evidence of EFIDA phenomena $C = f(U)$.

SUMMARY OF MAGNETIC FLUID CONTROLLED CAPACITOR

Interesting phenomena in magnetic fluids, MFIDA, res. EFIDA, during which the isotropic magnetic fluid under the influence of the external magnetic, resp. electric field, changes to anisotropic medium were highlighted. First of this phenomena was used to design a capacitor, whose capacity can be changed with the use of an external magnetic field.

The external magnetic field induces structural changes in the magnetic fluid and because of the weight of the ferromagnetic particles chains are time responses (i.e., capacity change, token of hysteresis) relatively slow, in order up to minutes. In addition, another attribute of magnetic liquids is the instability of their physical properties (e.g., effect of the ambient temperature, hydroscopicity, evaporation, and other) and with it the low accuracy in reproducing investigated phenomena, which apparently lowers the possible applicability of the designed controlled capacitor, especially e.g., for measurement purposes. The authors were not able to perform measurements on the capacitor for different types of magnetic fluids, but because of nowadays wide range of these materials, the existence of such a fluid, that exhibits even stronger capacity change and sensitivity, than in experiments described. Despite listed limitations we assume, that this, by our opinion innovative application of magnetic fluids, deepens the existing knowledge of the magnetic fluids and the designed controlled capacitor posses advantageous properties as well, e.g., simplicity and with it connected reliability, low price etc. and it will find its practical uses, for example in sensors detecting the changes of the magnetic field.

APPLICATIONS: SOME EXAMPLES OF MACHINERY USING FERROFLUID

Ferrofluids have found many useful applications in the design of some machine parts. Of many applications, we discuss rotary shaft seals and controlled shock absorbers.

Ferrohydrodynamic Sealing of Rotating Shafts

Figure 23 shows the principle of magnetic conductive shaft sealing.

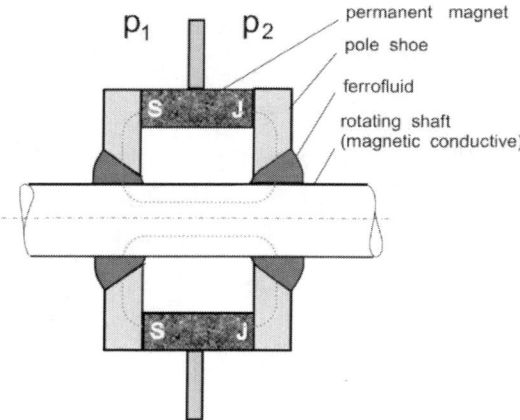

Figure 23. Principle of solution of rotating magnetically conductive shaft by ferro-liquid.

In separate spaces there are different pressures p_1 and p_2 of the gaseous medium. An example would be rotating electrical machines in which the rotor and its windings are cooled by a well thermally conductive medium, such as hydrogen. The permanent magnet in the form of a short hollow cylinder magnetized in the direction of the axis is supported by pole pieces made of a magnetically soft ferromagnetic material. There is an air gap of several tenths of a millimeter between the shaft and the pole pieces. Between the pole pieces and the shaft, the magnetic field is concentrated in a small space of the air gap. If ferrofluid is introduced into this area, it will be fixed by a magnetic field. With a magnetic induction in the air gap of about 1 T, the ferro-liquid seal maintains the pressure difference $|p_1 - p_2| \approx 0{,}2$ to 1 bar. This method of shaft seal is also used at the same pressures $p_1 = p_2$ as a very effective dust seal, for example to protect (and thus extend the service life) of bearings working in dusty or chemically aggressive environments. The ferrohydrodynamic seal is limited by the shaft diameter (used up to a shaft diameter of about 100 mm) and the number of revolutions (up to 6,000 rpm).

It is reported that in a conventional design, DN values of up to 200,000 can be achieved (where D is the shaft diameter in mm and N is its RPM); however, values of up to $DN = 500{,}000$ have been achieved. At high DNs, the ferrofluid should be cooled, as a high temperature would shorten its life. By constructional modification of the indicated principle, a multistage seal can be created, whereby the overpressure value $|p_1 - p_2|$ up to about 10 bar. Figure 24 shows the principle of a seal for a shaft that is magnetically nonconductive.

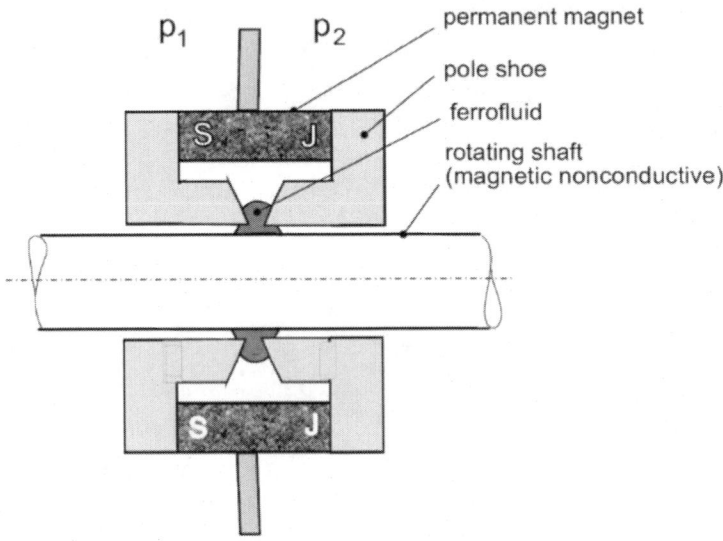

Figure 24. Principle of solution of rotating magnetically non-conductive shaft by ferrofluid.

Compared to conventional mechanical seals, the described method has several advantages: it is simpler (hence cheaper), more reliable, has less frictional momentum (ferrofluid also acts as a lubricant), high tightness, long service life (manufacturers say 10 to 15 years) and can work in a wide temperature range from −100 to 200°C. The application is versatile, e.g., for hard disk drives, vacuum pumps, vacuum grommets, compression refrigeration equipment, etc. It has also been proven for aircraft and astronomical instruments operating at high altitudes, manipulators and robots working with toxic or biologically active material.

Ferrohydrodynamic Dampers

Ferrohydraulic dampers dissipate the kinetic energy of undesirable deflections or oscillations of machine parts to thermal energy. Conventional hydraulic dampers exhibit constant damping during their operation, or damping can be varied by regulating the oil flow through a throttle valve that is mechanically adjusted. In shock absorbers with a flux, the current signal applied to the coil of the damper induces a magnetic field that increases the viscosity of the flux, thereby increasing damping. Figure 25 shows the design principles of ferrofluid shock absorber for linear movement. Magnetic shock absorbers have been successfully used in a variety of devices, from fine measuring instruments, automatic washing machines, truck seats to damping the effects of inequalities in the travel path of vehicles.

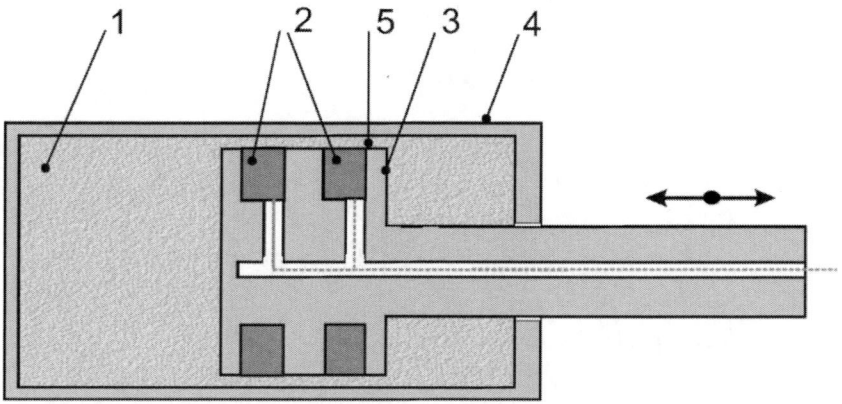

Figure 25. 1 ... ferrofluid, 2 ... powering coil, 3 ... shock absorber piston, 4 ... shock absorber housing, 5 ... gap

In cars, ferrohydrodynamic shock absorbers are very promising. If the car moves at a speed of 72 km/h, it will travel 2 cm in 1 ms. Conventional shock absorbers react for about 15 ms and the car will travel a 30 cm path before the shock absorber responds. On the other hand, the magnetic damper reacts significantly faster in about 5 ms. In this case, the shock absorbers react as soon as the 10 cm track has been driven and the car is carried over the uneven terrain without swinging. Improved damping prevents the transmission of vibration to the cab, reduces wheel bounce and thus the loss of adhesion between the tire and the road (among other things, reduces the likelihood of aquaplaning in the rain) and increases the stability of the car, especially when cornering. Magnetic shock absorbers therefore increase safety and comfort, shorten the braking response, improve the performance of the car and extend its service life, especially of tires. Bose proposes even more efficient car shock absorbers; they are designed as linear motors, which are controlled by a signal from the control unit, for each wheel separately. In this embodiment, optimum damping is achieved in as little as 1 ms.

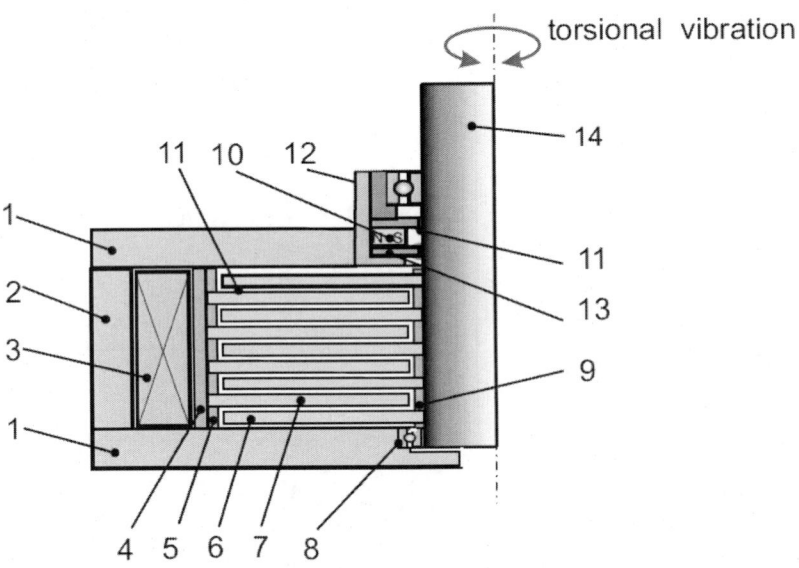

Figure 26. Controlled magnetorheological torsional damper.

Magnetic clutches are based on similar principle, when transmitting the torque between two shafts, Figure 27. *Magnetic break works on similar principle*. See also [1, 19, 20, 21, 22].

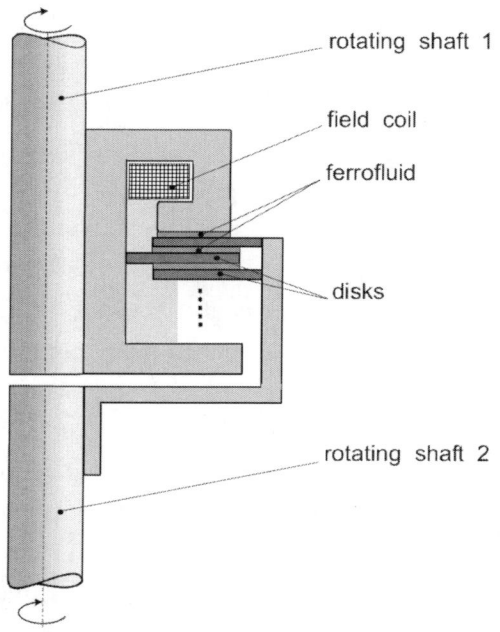

Figure 27. Controlled magnetorheological clutch.

APPLICATION: PERISTALTIC PUMP WITH MAGNETOELASTIC DRIVE

Classic liquid pumps are based on different physical principles. These are piston, diaphragm, centrifugal, and tooth pumps. Peristaltic pumps are a special category (see [15, 23, 29]). Generally, peristalsis can be understood as a fluid transport mechanism that occurs when a flowing force wave acts on the elastic tube with the pumped liquid. Peristaltic pumps induce flow within a fluid filled flexible tube through forward traveling wall contraction. These are available in a variety of configurations. The operation of one of the types of peristaltic pumps is shown in Figure 28. The peristaltic pumps are bidirectional.

The advantage of peristaltic pumps is that it does not contain valves and seals, which facilitate their maintenance and reduces their cost. While conventional pumps are pumped liquid in contact with the metal parts of the pump, the metal ions pass into the pumped liquid and contaminate it, but it is not so for peristaltic pumps. Therefore, they can be effectively used especially in the medical and pharmaceutical industries, where they meet strict conditions of aseptic regime. The liquid pumped by the peristaltic pumps need not be homogeneous and may contain small solid particles. An example is the pumping of blood and their derivatives by peristaltic pumps that behave gently toward the blood elements. In conventional pumps, the use of chemically abrasive or aggressive liquids leads to corrosion of the valves and thus to their rapid wear, i.e., to their limited lifetime and lower reliability. Electric motors are used to propel conventional pumps, which, together with pump operation, are a source of noise. The drive of the peristaltic pump described below does not use movable mechanical components and is therefore silent. As mentioned above, not only peristaltic pumps are used mainly in medicine (e.g., as a peritoneal dialysis pump, a insulin pump for diabetics, and a micropump of artificial heart), in the pharmacological and chemical industry and chromatography, but also as pumps for waste water pumping, or pumps for pumping concrete mixtures.

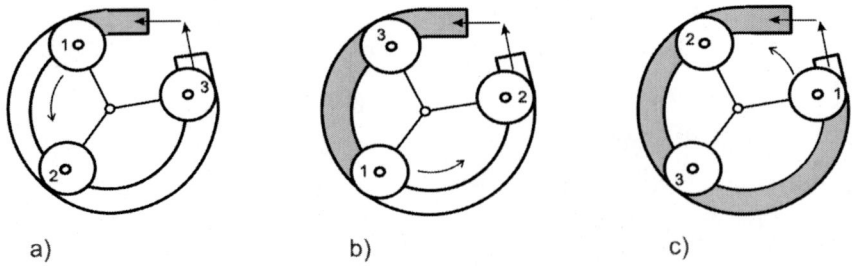

Figure 28. Principle of a peristaltic pump with motor drive.

The peristaltic pump described further differs from the others by its drive providing a running magnetic wave, which is generated by multiphase windings. The force action of this magnetic wave on the pumped liquid mediates a magnetoelastic medium that is present either in liquid form (i.e., as magnetic fluid, [28]) or in solid form (e.g., as a magnetic rubber).

Design of a Peristaltic Pump

The described peristaltic pump can be arranged as linear or circular.

Linear Arrangement of the Peristaltic Pump

The two blocks composed of the dynamic plates are placed above each other (Figure 29). Two chambers are inserted between them; are made of non-ferrous material and the area on which the two chambers meet, forms an elastic membrane. The upper chamber has attached suction and discharge pipes at its ends and pumped liquid. The lower chamber is closed and is filled with magnetic fluid. Pump drive provides a running magnetic wave that moves between the ferromagnetic blocks. The top cube is provided with grooves into which a two-phase (generally multiphase) winding is inserted, similar to a linear asynchronous or synchronous motor stator. The winding is fed by a two-phase (or multiphase) current, inducing a magnetic field $B(t, v)$ that is in the form of a magnetic wave that moves at speed v. The velocity v is dependent on the frequency of the two-phase multi-phase excitation current. The running magnetic wave acts in the space of the two chambers (see Figure 30). It invokes the rise of the magnetic fluid, which by force P gradually mechanically compresses the membrane of the chamber with the pumped liquid. Running magnetic waves cause the two membranes to move inward, and at one end of the pumped liquid chamber a suction effect occurs and a displacement effect occurs on the opposite chamber of the vessel. Magnetic wave velocity and therefore the linear velocity of the membrane lift is

$$v = \frac{f \cdot l}{p} \quad [m/s] \tag{28}$$

where f [Hz] is the frequency of the two-phase (multiphase) current in the winding, l is the linear pump length, and p is the number of pole pairs for which the multiphase winding is made. While rotating electric machines typically use a frequency $f = 60$ Hz, in the peristaltic pump, because of the magnetic viscosity of the magnetic fluid, it is necessary to use a substantially lower frequency, in the order of hertz units. For example, with a two-pole winding ($p = 1$) and a frequency $f = 1$ Hz, the magnetic wave and hence the lengthened part of the ferrofluid moves at $v = l$ [m/s]. Thus, the velocity of the feed stream can regulate the velocity in (and hence the amount) of pumped liquid (see Figure 31).

From the velocity v and volume V [m³] of the pumped liquid chamber, we determine the volume of pumped liquid per unit time

$$Q = \frac{v \cdot V}{l} \quad \left[m^3 / s \right] \tag{29}$$

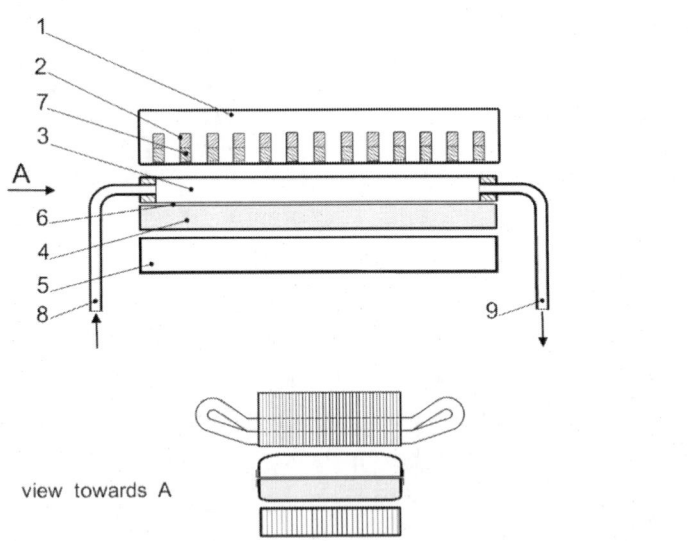

Figure 29. Linear arrangement of the peristaltic pump without the wave of magnetic field. 1 —block from dynamo sheets whose winding generates a running wave, 2 — slots in which two- or multi-phase windings are inserted, 3 — chamber with a pumped liquid, 4 — closed channel with magnetic fluid, 5 — joke from dynamo sheets, 6 — elastic membranes, and 7 — two- or multi-phase winding.

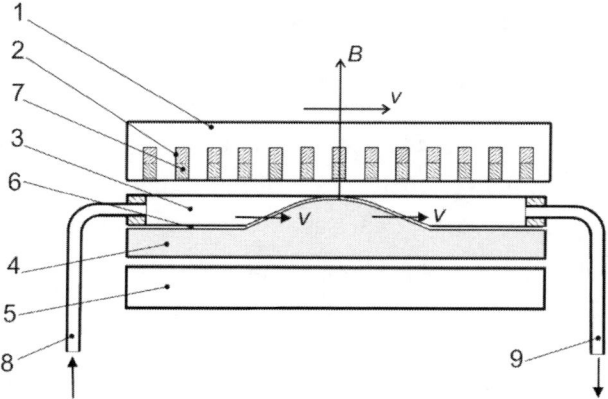

Figure 30. Linear arrangement of the peristaltic pump under the action of the running magnetic wave. 1 to 6 as in the previous Figure 29, 7 — front of the wave of the magnetic fluid pumped, and 8 — tale of the wave of the magnetic fluid suction of liquid.

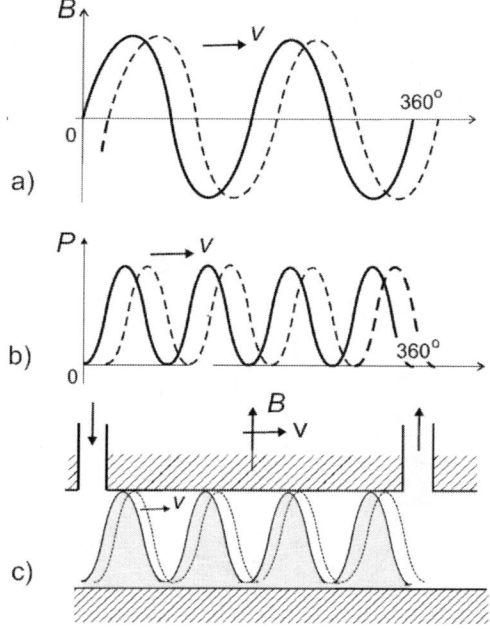

Figure 31. (a) Running magnetic wave. (b) Density of force acting on the magnetic fluid. (c) Movement of the bulge of the membranes separating the magnetic fluid and pumped liquid. (B is the magnetic flow density, P is the density of force, and v is the velocity of running magnetic wave).

Circular Arrangement of the Peristaltic Pump

This arrangement is based on the same principle, but in a cylindrical configuration. The blocks made up of dynamic sheets are replaced by two rings (Figure 32). A pair of chambers having the shape of concentric rings and separated by elastic membranes is inserted into the space between the two rings. The liquid-filled chamber is provided with suction and displacement line, the chamber filled with magnetic fluid is closed. The outer ring has spots on its inner surface with a two-phase (or multiphase) winding, which is made similar as the stator winding of the induction or, synchronous electrical machine.

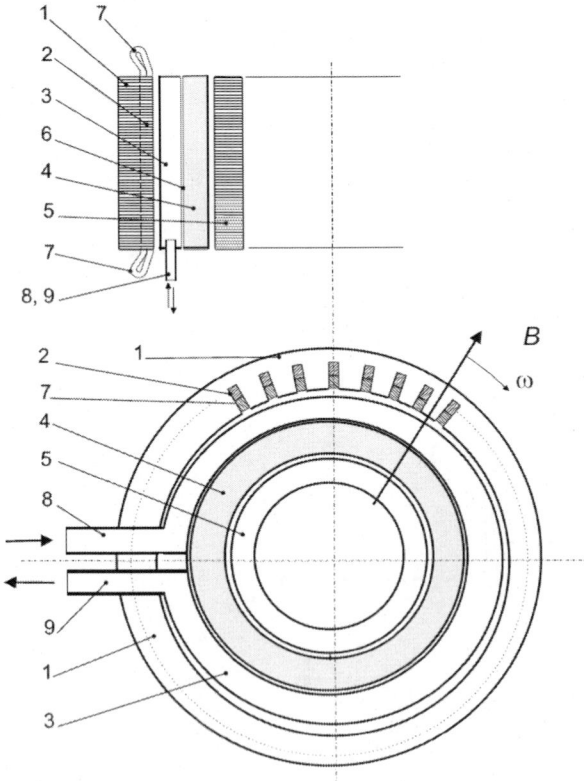

Figure 32. Circular arrangement of the peristaltic pump. 1 to 6 as in the previous Figure 29.

The two-phase (or multiphase) current in the winding induces a rotating magnetic field in air gap between fixes and movable parts of the pump. The wave of the magnetic field causes the magnetic fluid to emit, which is transmitted through the elastic membranes to the chamber with the pumped liquid, and its rotation causes suction and displacement of pumped liquid. The rotational speed of the rotating magnetic field is

$$n = \frac{f}{p} \quad [\text{rps}]$$

(30)

From eq. above, we can easily determine the amount of liquid pumped per unit of time.

Peristaltic Pump Summary

Described peristaltic pump can be used especially in special applications, which use its some of the following advantageous properties.

Pumped liquid are contactless between pump components. There is no damage to any microparticles contained in the pumped liquid (e.g., blood cells when pumping blood).

The liquid sterility requirement can be quite easily fulfilled, the pumped liquid being chemically aggressive, possibly toxic, or bacteriologically active. The pumped liquid need not be homogeneous and may contain solid particles.

The operation of the peristaltic pump can be continuous, the amount of pumped liquid can be continuously regulated, i.e., by the frequency of the supply current. The pump can also be used to accurately (e.g., medicate) dispense larger volumes of liquid. Pump operation can be computer controlled, the pump can be connected to the programmed system.

The operation of the peristaltic pump is relatively silent and maintenance free; the pump is almost trouble-free, and its service life is long-lasting. Replacing the used liquid chamber behind the new (aseptic) chamber is easy and non-expensive.

The initial cost and operating cost of this peristaltic pump are low.

The drawbacks include the pump requires a two-phase (or multiphase) power supply with a low output frequency (in the order of hertz) or with an adjustable frequency if the pump's drive is controlled by the frequency.

CONCLUSION

Ferrofluids are characterized by their anisotropy and by the change of physical parameters, especially viscosity, dielectric permittivity and magnetic permeability under the effect of external magnetic field. Several examples from machinery practice using these properties were described. The possible applications of magnetic fluids are however much wider. Ferrofluids are used in the design of special electrical machines, transformer and actuators, in acoustic to design loudspeakers and even in biomedicine. Ferrofluids are relatively new and yet not well known material. They can be expected to become an even more important part of material engineering and play an important role in the development of new technologies.

ACKNOWLEDGMENTS

This text was created under the support of University of West Bohemia, SGS-2018-043.

REFERENCES

[1] Rabinow, J. Magnetic fluid clutch. *Nat. Bur. of Stand. Techn. News Bull.,* 32, no. 4, (1948), pp. 54-60.

[2] Berkovski, B. and Bashtovoy, V. (eds.). *Magnetic fluids and application Handbook.* New York, Begell House, Inc., publishers, 1996.

[3] Berkowski, B. M. and Medvedev, V. F. and Krakov, M. S. *Magnetic Fluids, Engineering Applications.* Oxford University Press, Oxford, New York, Tokyo, 1993.

[4] Blums, E. and Cebers, A. and Maiorov, M. M. *Magnetic Fluids.* Walter de Gruyter, Berlin, 1997.

[5] Franklin, Thomas A. *Ferrofluid Flow Phenomena.* Thesis, Massachusetts Institute of Technology, June 2003.

[6] Mayer, D. Future of electrotechnics: Ferrofluids. *Advances in Electrical and Electronic Engineering*, Vol. 7, no. 1-2 (2008), pp. 9-14.

[7] Odenbach, S. (editor) *Colloidal magnetic fluids.* Springer-Verlag, Berlin, 2009.

[8] Oldenbach, S. *Magnetoviscous effects in ferrofluids.* Springer-Verlag, Berlin 2002.

[9] Oldenbach, S. (editor) *Ferrofluids.* Springer-Verlag, Berlin, 2002.

[10] Rosensweig, R. E. Fluid Dynamics and Science of Magnetic Liquids. *Advances in Electronics and Elektron Physics*, vol. 48, (1979), pp. 103-199.

[11] Rosensweig, R. E. *Ferrohydrodynamics.* Dower Publications, Inc., Mineola, New York, 1997.

[12] Butter, K. and Bomans, P. H. H. and Frederik, P. M. and Vroege, G. J. and Philipsa, A. P. Direct observation of dipolar chains in iron ferrofluids by cryogenic electron microscopy. *Nature Materials*, vol. 2, Feb. (2003), pp. 88-91.

[13] Mayer, D. An Approach to Measurement of Permeability/ Permittivity Tensor of Ferrofluids. *Journal of El. Eng.*, 66 no. 5 (2015), pp. 292–296.

[14] Mayer, D. and Polcar, P. A novel approach to measurement of permeability of magnetic fluids. *Przeglad Elektrotechniczny,* 2012, y. 88, vol. 7B, pp. 229-231. ISSN: 0033-2097.

[15] Krzeminski, S. K. et al. Numerical analysis of peristaltic MHD flows. *IEEE Transaction on Magn.*, vol. 36, no. 4, Jul. 2000, pp. 1319–1324.

[16] Chantrell, R. W. and Bradbury, A. and Popplevell, J. and Charles, S. W. Aglomeration formation in magnetic fluid. *J. Appl. Phys.*, 53, no. 3, pp. 2742-2744.

[17] Byrne, J. V.: Energy-Based Formulation of the Stresses in Nonlinear Polarizable Fluids. *IEEE Trans. on Magnetics*, Vol. MAG-10, No. 2, June 1974, pp. 358-361.

[18] Byrne, J. V. Ferrofluid hydrostatice according to clasical and recent theories of the stresses. *Proc. IEE*, vol. 124, no. 11, Nov. 1977, pp. 1089-1097.

[19] Jaindl, M. et al. Optimal design of a disk type magneto-rheological fluid clutch. *Przeglad elektrotechniczny*, 83, no. 6, (2007), pp. 25-29.

[20] Lampe, D. Anwendung von magnetorheologischen Fluiden in Kupplungen. *Antriebstechnik*, 38 (1999), Nr. 7, str. 59-62.

[21] Szelag, W. *Finite element analysis of the Magnetorheological fluid brake transients*. Compel, vol. 23, no. 3, (2004), pp. 758-766.

[22] Polcar, P. Design, construction and experimental verification of magnetorheological brake. *Communications - Scientific Letters of the University of Zilina*, 2013, y. 15, vol. 2A, pp. 23-27. ISSN: 1335-4205.

[23] Burns, J. C. and Parkes, T. Peristaltic motion. *J. Fluid Mech.*, vol. 29, no. 4, (1967), pp. 731–743.

[24] Kim, W. G. and Oh, J. G. and Choi, B. A study on the development of a continuous peristaltic micropump using magnetic fluids, Sens. *Actuators A, Phys.*, vol. 128, no. 1, (2006) pp. 43–51.

[25] Klespitz, J. and Kovacs, L. Peristaltic pumps—a review on working and control possibilities. *Proc. IEEE 12th Int. Symp., Appl. Mach. Intell. Inform.* (SAMI), Herlany, Slovakia, Jan. 2014, pp. 191–194.

[26] Kumar, B. V. R and Naidu, K. B. A numerical study of peristaltic flows. *Comput. Fluids*, vol. 24, no. 2, (1995), pp. 161–176.

[27] Neto, A. G. et al. Linear peristaltic pump driven by three magnetic actuators: simulation and experimental results. Proc. *IEEE Instrum. Meas. Technol. Conf.* (I2MTC), May 2011, pp. 1–6.
[28] Stork M. and Mayer, D. Peristaltic pump with magnetoelastic drive. *IEEE Transactions on Magnetics*, (2018).
[29] Sud, V. K. and Sekhon, G. S. and Mishra R. K. Pumping action on blood by a magnetic field. *Bull. Math. Biol*, vol. 39, no. 3, (1977), pp. 385–390.
[30] Vijayaraj, K. and Krishnaiah, G. and Ravikumar, M. M. Peristaltic pumping of as fluid of variable viscosity in a non-uniform tube with permeable wall. *Journ. Theor. Appl. Inf. Technol.*, vol. 8, no. 1, (2009), pp. 82–91.
[31] Polcar, P. and Mayer, D. Magnetic field controlled capacitor. *Journal of Electrical Engineering,* 2016, y. 67, vol. 3, pp. 227-230. ISSN: 1335-3632.
[32] Cecelja, F. and Rakowski, R. and Martinez, L. *A Ferrofluid-based Magnetic Field Sensor. Instrum. and Meas. Technology Conference*, IMTC 2005, Otawa, Canada, May 2005, pp. 612-616.
[33] Vislovich, A. and Polevikov, V. Effect of the centrifugal and capillary forces on the free surface shape of a magnetic liquid seal. *Magnetohydrodynamics,* 30 (1994), pp. 67-74.
[34] Polcar, P. and Mayer, D. Investigation of field induced magnetic anisotropy in ferromagnetic liquids. In 11th International Conference Elektro 2016 proceedings. *Piscataway: IEEE,* 2016. pp. 565-568. ISBN: 978-1-4673-8698-2.

INDEX

A

aggregations, 91
anisotropic, viii, ix, 5, 41, 48, 49, 81, 90, 91, 93, 95, 100, 101, 102, 103, 104, 105, 106, 107, 111, 119, 123
anisotropic environment, 91, 95, 100, 101, 103, 106
anisotropic medium, ix, 90, 103, 111, 123
anisotropy, v, vii, viii, ix, 1, 2, 3, 4, 5, 6, 7, 10, 12, 14, 15, 16, 17, 20, 23, 25, 26, 27, 28, 29, 31, 32, 33, 34, 35, 36, 37, 38, 39, 40, 41, 44, 45, 48, 49, 55, 56, 58, 59, 61, 62, 63, 67, 68, 69, 71, 73, 74, 76, 83, 84, 85, 86, 90, 100, 109, 111, 112, 113, 114, 118, 119, 120, 136, 139
atmosphere, 52, 64, 65, 66, 78, 83
atoms, 49, 67, 84

B

basic equation, 106
Bernoulli equation, 108
bias, 38, 70, 87

C

blood, 130, 135, 139
boundary value problem, 106

capillary pressure, 108
carbon, vii, 1, 2, 3, 20, 29, 37, 38, 40, 41, 42, 43
carbon nanotubes, v, vii, 1, 2, 3, 29, 35, 37, 38, 40, 41, 42, 43, 44, 45
carrier liquid, 91, 92, 93, 110, 120
carrier medium, 92, 112
chemical, vii, ix, 1, 3, 8, 37, 40, 52, 89, 91, 92, 93, 94, 112, 130
chemical industry, 130
chemical properties, ix, 89
chemical vapor deposition, vii, 1, 3, 37
chemical vapour deposition, 40
Circular Arrangement of the Peristaltic Pump, 134
composite pressure, 108
composition, viii, 48, 49, 52, 65, 85, 94
conductivity, 17, 18, 100, 105, 106, 107
conductivity tensor, 105, 106
configuration, 34, 35, 113, 115

core-shell structure, 48
correlation, viii, 1, 2, 4, 5, 6, 7, 13, 14, 15, 16, 23, 45, 49, 69, 72
correlation function, viii, 1, 2, 6, 7, 13, 14, 15, 16, 23
correlation function of the magnetic anisotropy axes, 2, 23
correlations, 5, 15, 17, 23, 36
crystalline, viii, 5, 21, 33, 48, 54, 56, 64, 66, 79, 80, 81
crystallinity, ix, 48, 73, 80, 84
crystallites, 54, 56, 74
crystallization, 74
curie temperature, 23, 92
current density, 107

D

defects, 17, 21, 81, 84
demagnetization, 24, 59, 77, 93
destabilization of ferrofluid, 92
destruction, ix, 48, 81, 84
detergent, 91, 92, 94
deterioration, 92
deviation, 50, 55, 63, 75, 78
dielectric permittivity, ix, 89, 90, 136
diffraction, 52, 54, 64, 74, 79, 80, 86
dimensionality, 5, 6, 10, 12, 15, 23, 31
disorder, 39, 42, 64
dispersion, 31, 55, 57, 58, 67, 69, 78, 80, 84
displacement, 131, 134
dissolved oxygen, 81, 84
distribution, 9, 20, 22, 24, 25, 26, 27, 28, 35, 50, 58, 67, 75, 86, 93, 106, 111, 123
distribution function, 25, 26, 27, 28
dynamics of magnetic viscosity phenomenon, 108

E

electric current, 97
electric field, 42, 53, 104, 105, 106, 112, 113, 120, 123
electric field induced dielectric anisotropy, 120
electrical breakdown, 42
electrodes, 120, 121
electromagnetic, ix, 90, 100, 103, 106
electron, 8, 51, 52, 110, 137
electron microscopy, 8, 110, 137
element method, 116
energy, ix, 2, 20, 32, 42, 51, 52, 61, 62, 63, 90, 111, 127
environment, 17, 90, 91, 94, 95, 101, 103, 106
examination of the permeability tensor, 115
excitation, 112, 113, 114, 115, 116, 118, 131
external environment, 29
external magnetic fields, 111

F

ferrofluid, vii, x, 90, 93, 95, 96, 97, 98, 99, 103, 106, 107, 111, 112, 113, 114, 115, 117, 118, 124, 125, 126, 127, 132, 137, 138, 139
ferrohydraulic dampers, 127
ferromagnetic, vii, viii, 1, 2, 3, 4, 5, 8, 9, 10, 17, 19, 29, 31, 35, 36, 38, 39, 40, 41, 44, 48, 50, 52, 55, 56, 58, 60, 63, 64, 69, 70, 73, 74, 75, 78, 81, 82, 83, 84, 86, 90, 91, 92, 119, 120, 121, 123, 125, 131, 139
ferromagnetic liquid, 90, 139
ferromagnetism, 39
ferromagnets, 12, 29, 39, 40, 41, 92
field induced magnetic, 109, 111, 139
films, vii, viii, 40, 44, 45, 48, 49, 50, 51, 52, 53, 54, 55, 56, 57, 58, 59, 63, 64, 66, 69, 70, 72, 73, 74, 76, 77, 78, 79, 80, 81, 82, 83, 84, 85, 86
finite, 116, 138

finite element method, 116
fluctuations, 28, 39, 57, 67, 71, 76, 77
fluid, ix, 90, 95, 108, 119, 120, 121, 122, 123, 136, 138
fluid-magnetic pressure, 108
force, 26, 29, 93, 95, 98, 100, 101, 119, 129, 130, 131, 133
formation, vii, viii, ix, 2, 3, 4, 8, 48, 49, 50, 64, 66, 69, 71, 78, 82, 138

G

graphene sheet, 39, 42
graphite, 42, 52
gravitational force, 95, 98
growth, 8, 10, 37, 49, 55, 56, 58, 67, 77, 79, 80, 84

H

hysteresis, 4, 22, 30, 70, 93, 111, 122, 124
hysteresis loop, 4, 22, 30, 70

I

image, 8, 9, 10, 20, 29, 30, 64, 66, 72, 73, 74, 75, 78
induced anisotropy, 110, 112, 113, 118
induced dielectric anisotropy, 109, 112, 119
induced dynamic anisotropy, 118
induction, 103, 105, 120, 125, 134
instability, 90, 95, 124
interaction effect, 44
interaction effects, 44
ion irradiation, v, 47, 48, 86
ions, viii, 48, 51, 52, 67, 72, 73, 75, 77, 78, 79, 82, 83, 84
iron, 8, 38, 55, 65, 67, 92, 111, 137
irradiation, viii, 48, 51, 72, 73, 74, 75, 76, 78, 79, 80, 81, 83, 85, 86

isotropic, 54, 93, 100, 101, 102, 103, 106, 107, 108, 111, 118, 119, 123
isotropic medium, 100, 111, 119
isotropic to anisotropic state, 118

L

law of the approach to magnetic saturation, 2, 4
levitation, x, 90, 98
linear arrangement of the peristaltic pump, 132, 133
linear dependence, 27
liquids, vii, x, 90, 93, 94, 119, 120, 124, 130
losses, ix, 90, 93
low temperatures, viii, 2, 4, 14, 67, 70, 74, 75

M

magnet, 95, 98, 99, 100, 125
magnetic anisotropy, 1, iii, v, vii, viii, 1, 2, 3, 4, 5, 6, 14, 17, 19, 20, 23, 25, 29, 31, 35, 36, 38, 40, 45, 48, 55, 58, 59, 61, 63, 67, 69, 71, 73, 74, 76, 83, 85, 86, 89, 90, 111, 112, 115
magnetic break, 128
magnetic characteristics, 35
magnetic clutches, 128
magnetic conductive shaft sealing, 124
magnetic dipole, 29, 91
magnetic field, ix, 6, 7, 10, 19, 20, 24, 26, 30, 33, 53, 55, 57, 58, 59, 60, 62, 76, 89, 90, 91, 93, 94, 95, 96, 98, 99, 105, 106, 107, 108, 109, 110, 111, 112, 113, 114, 115, 116, 117, 118, 119, 120, 122, 123, 125, 127, 136, 139
magnetic field controlled capacitor, 119, 120, 139
magnetic field effect, 120
magnetic fields, 6, 53, 91, 115

magnetic induced anisotropy, 119
magnetic materials, 86
magnetic moment, ix, 2, 20, 22, 36, 48, 49, 53, 55, 56, 57, 59, 60, 61, 67, 69, 70, 73, 76, 78, 80, 81, 84, 91, 93, 94
magnetic normal, 108
magnetic particles, 39
magnetic properties, 2, 4, 12, 15, 35, 41, 43, 44, 69, 72, 73, 75, 83, 111
magnetic relaxation, viii, 2, 24, 36, 43
magnetic resonance, 49
magnetic resonance spectroscopy, 49
magnetic rheological, 90
magnetic structure, 35, 49, 86
magnetic viscosity, 24, 25, 26, 27, 28, 36, 90, 93, 94, 132
magnetically/mechanical fields, 107
magnetism, 38, 41
magnetization, ix, 2, 4, 5, 6, 20, 24, 25, 27, 30, 31, 34, 35, 39, 41, 44, 53, 57, 58, 59, 60, 62, 63, 68, 70, 74, 76, 77, 90, 91, 92
magnetoelastic anisotropy, viii, 2, 31, 32, 33, 34
magnetoelastic medium, 130
magnetorheological effect, 119
magnetorheological liquids, 94
magnetorheology, 90
magnetorheoloigical liquids, 93
magnetostriction, 33
magnetoviscous effect, 109, 137
magnets, 39, 43, 95
materials, 3, 4, 5, 21, 29, 37, 49, 91, 92, 124
mathematical description, 100, 107
matrix, 2, 3, 4, 17, 20, 32, 40, 49, 53, 55, 64, 74, 85, 101, 103, 104, 111
matter, iv
measurement, ix, 22, 29, 42, 70, 90, 109, 124, 137
measurements, viii, 2, 6, 24, 41, 53, 122, 124
measuring the permeability tensor, 114, 118
measuring the permittivity, 117

mechanical properties, 101
mechanic-elastic qualities, 93
membranes, 131, 132, 133, 134
microparticles, 93, 135
microscope, 51, 52, 110
microstructure, 53, 110
microstructure of the ferrofluids, 110
multilayer planar capacitor, 121
multiwalled carbon nanotubes, 43, 44, 45

N

nanocomposites, viii, 2, 3, 4, 6, 12, 31, 35, 37, 48, 50, 53, 64
nanocrystals, 44
nanometers, viii, 1, 17, 18, 19, 49
nanoparticles, v, vii, viii, 1, 2, 3, 8, 25, 29, 32, 35, 36, 37, 38, 40, 41, 42, 44, 47, 48, 49, 51, 54, 58, 63, 70, 83, 85, 86, 87, 91, 92, 93, 94, 110, 115, 118
nanoribbons, 38, 39
nanostructured materials, 3
nanostructures, 41, 49, 50, 53, 54, 55, 58, 64, 70, 74, 87
nanotechnologies, ix, 89
nanotube, vii, viii, 2, 3, 8, 9, 10, 19, 35, 38, 40, 42, 44
nanowires, 3, 5, 32, 37, 43, 44, 45
Navier-Stokes equations, 108
nonhomogenous magnetic field, 95

O

oxidation, v, viii, 47, 48, 50, 51, 52, 64, 66, 67, 68, 69, 70, 76, 78, 80, 83, 84, 85
oxygen, viii, 48, 51, 52, 64, 65, 66, 78, 83, 84

P

parallel, 10, 14, 30, 31, 32, 33, 34, 53, 104, 116
particle mass density, 108
passive levitation, 98
peristaltic pumps, vii, 129, 138
permanent magnet levitation, 98
permeability, ix, 31, 89, 90, 92, 100, 105, 107, 109, 110, 111, 113, 114, 115, 116, 118, 136, 137
permittivity, ix, 89, 90, 100, 104, 109, 110, 112, 117, 118, 119, 120, 136, 137
permittivity tensor, 104, 109, 112, 117, 118, 119, 120, 137
perpendicular magnetic anisotropy, v, vii, viii, 47, 48, 49, 83, 86
physical properties, ix, 89, 90, 110, 111, 112, 119, 124
physical structure, 100
physical theories, 91
plastic deformation, 3
polar molecules, 91
polarization, 114, 115, 118
pressure, viii, 48, 51, 108, 125
pressure in nonmagnetic fluid, 108
production technology, 90
pumps, vii, x, 90, 126, 129, 138

R

Raman spectra, 21, 42
Raman spectroscopy, 20, 43
relaxation, viii, 2, 4, 20, 24, 25, 27, 28, 36, 43
relaxation model, 28
relaxation process, 20, 24, 27
relaxation processes, 20
response, 109, 111, 122, 123, 128
running magnetic wave, 130, 131, 133

S

saturation, 2, 4, 6, 23, 25, 26, 27, 30, 38, 39, 41, 58, 59, 61, 68, 69, 70, 92, 93, 120
scalar potential, 106, 107
scalars, 100, 103, 104
sediment, 91, 93, 94
sedimentation, 90, 94
self-inductance coil test, 113
shape, vii, viii, 2, 4, 10, 30, 31, 48, 49, 50, 52, 53, 55, 56, 58, 59, 60, 61, 62, 63, 64, 69, 73, 75, 76, 77, 83, 84, 86, 96, 113, 116, 134, 139
simpler experiments, 94
solid ferromagnetics, 93
solution, 91, 94, 116, 125, 126
spectroscopy, viii, 21, 42, 48, 50, 51, 52, 53, 55, 56, 58, 61, 67, 69, 73, 74, 75, 80, 81, 82, 83, 84, 86
square matrix, 101
state, viii, 12, 32, 48, 71, 73, 84, 91, 93, 108, 109, 111, 112, 113, 114, 115, 117, 118, 122
structural changes, 76, 111, 119, 122, 123
structural transformations, 73
structure, viii, 10, 20, 36, 48, 49, 54, 64, 67, 79, 80, 83, 85, 91, 120
superparamagnetic, 49, 55, 69, 76, 81
system of spikes, 95

T

temperature, viii, 2, 4, 10, 11, 12, 15, 22, 23, 24, 25, 26, 27, 28, 36, 39, 41, 53, 57, 70, 74, 76, 92, 112, 124, 126
temperature dependence, viii, 2, 15, 27, 36
tensor, ix, 90, 101, 103, 104, 105, 109, 111, 113, 115
thermal (Brown) motion, 93
thermal decomposition, 3
thermal energy, 2, 15, 127

thermal stability, 2
thermodynamic, 108
traction, 108
transformation, 52, 69, 78, 107
transmission, viii, 48, 49, 52, 53, 128
transmission electron microscopy, viii, 48, 49, 52
treatment, vii, viii, 8, 48, 50, 51, 72, 76, 77, 81, 83, 84, 112

V

vector, 103, 104, 105, 106, 107, 111
vector potential, 106
velocity, 24, 131, 132, 133
viscosity, ix, 24, 25, 26, 27, 28, 36, 89, 90, 93, 94, 108, 110, 119, 120, 127, 132, 136, 139